Professional Practice for Quantity Surveyors in the 21st Century

This book covers developing trends and changes occurring in the quantity surveying profession. It begins by explaining the construction industry structure, followed by the quantity surveying practice modalities, professional ethics, roles of regulatory bodies, continuous professional development, and code of professional conduct of quantity surveying practice. Further topics covered include international construction, contract administration, financial management, and liquidated and ascertained damages as applicable to the construction industry.

Features:

- Considers the structure of the construction industry and the inter-relationship of professionals in the built environment.
- Provides a compendium of quantity surveying professional practice, including developing country perspectives and examples.
- Includes information on cultural differences and how they may impact quantity surveying practices, such as communication styles, decision-making processes, and work habits.
- Focuses on current industry trends and practices, such as sustainable practices, building information modeling (BIM), and advanced technologies.
- Covers financial statements and progress reports regarding construction projects.

This book is aimed at professionals and students in construction, quantity surveying/value management, and civil engineering.

Professional Practice for Quantity Surveyors in the 21st Century

Andrew Ebekozien and Clinton Aigbavboa

CRC Press is an imprint of the
Taylor & Francis Group, an **informa** business

First edition published 2025
by CRC Press
2385 NW Executive Center Drive, Suite 320, Boca Raton FL 33431

and by CRC Press
4 Park Square, Milton Park, Abingdon, Oxon, OX14 4RN

CRC Press is an imprint of Taylor & Francis Group, LLC

ISBN: 978-1-032-66143-8 (hbk)
ISBN: 978-1-032-60978-2 (pbk)
ISBN: 978-1-032-66139-1 (ebk)

DOI: 10.1201/9781032661391

Typeset in Times Lt Std
by Apex CoVantage, LLC

This book is dedicated to the Lord almighty God for the gift of life.

Contents

Foreword

The 21st-century construction industry faces unprecedented challenges and opportunities. The rapid pace of technological advancement, coupled with increasing environmental concerns and global economic pressures, has shown the critical need for a further understanding of construction economics. It is within this dynamic and complex landscape that the book *Principles of Basic Construction Economics in the 21st Century* emerges as an essential guide for navigating the future of construction. The motivation behind this book stems from a recognition of the evolving demands placed on construction professionals today. Despite the wealth of knowledge available, a gap exists between theoretical economic principles and their practical application in construction projects. This book seeks to bridge that divide, offering a foundation in construction economics while addressing the practical challenges of the industry.

The book is structured to provide a thorough understanding of the construction industry's economic underpinnings, quantity surveying practices, professional ethics, and other critical topics. Each chapter of the book builds upon the last to offer a holistic view of construction economics. In an era marked by sustainability concerns and the quest for efficiency, the principles outlined in this book are more relevant than ever. By examining the economic impact of the construction industry, the importance of ethical practices, and the principles of contract administration, the book provides readers with the tools to address today's challenges head-on. The book *Principles of Basic Construction Economics in the 21st Century* is written to serve a broad audience. For students, it offers a foundational text that bridges academic theory and practice. Professionals in the field will find it a valuable resource for updating their knowledge and refining their skills. Likewise, scholars and researchers will appreciate the book's contribution to the ongoing discourse in construction economics. Through real-world examples and case studies, the book illustrates the practical application of economic principles in construction financial management. It underscores the importance of sound financial management, effective contract administration, and the ethical considerations that underpin successful projects. Looking ahead, the book invites readers to consider the future evolution of construction economics. The book encourages ongoing education and adaptation to emerging trends, ensuring that the construction industry continues to thrive in the face of future challenges. The book *Principles of Basic Construction Economics in the 21st Century* is more than just a monograph; it is a call to action for all stakeholders involved in the construction industry. By offering a comprehensive overview of economic principles and their application, the book aims to equip readers with the knowledge and skills necessary to lead the industry forward.

Preface

In the dynamic landscape of the 21st-century construction industry, the principles of construction economics have never been more critical. This book is carefully thought through to serve as an essential guide for students, professionals, and scholars interested in the economic dimensions of construction. With the construction sector evolving rapidly due to technological advancements, regulatory changes, and global economic shifts, a deep understanding of its economic underpinnings is vital. The book begins with an in-depth exploration of the construction industry's nature, scope, and significant impact on national and international economies. Chapter 1 sets the stage by delineating the industry's characteristics, the pivotal role of key participants, and the overarching influence of regulatory and professional bodies. This foundational knowledge will equip readers to grasp the intricacies and unique challenges inherent in construction economics. The subsequent chapters delve into the heart of quantity surveying practice, professional ethics, tendering methods, contract administration, insurance, financial management, and the resolution of conflicts within construction contracts. Each chapter is written to build upon the last, thus ensuring a comprehensive understanding of both theoretical principles and practical applications. Our discussion on quantity surveying—spanning from its origin, roles, and ethical considerations to its critical function in tendering, procurement, and contract administration—underscores the profession's centrality in achieving economic efficiency and project success.

The book also addresses the multifaceted differences in quantity surveying practices across developed and developing countries, providing a global perspective on construction economics. Furthermore, it emphasizes the importance of professional ethics and continuous development, reflecting the commitment to fostering integrity and excellence in the construction profession. *Principles of Basic Construction Economics in the 21st Century* aims not only to inform but also to inspire the next generation of construction economics scholars. It is written to be a resource that readers will return to throughout their careers, whether they are just beginning in the field of construction economics or seeking to expand their expertise. Our dedication to this book reflects our belief in the power of thoughtful publication to elevate the construction profession and contribute to the development of sustainable, economically viable, and innovative construction practices worldwide. As the construction industry continues to serve as a cornerstone of economic development, the principles laid out in this book will empower professionals to navigate the complexities of modern construction projects. It is with great enthusiasm that we invite you to embark on this journey of discovery and professional growth.

Acknowledgements

We wish to express our profound gratitude to our families, mentors, and colleagues, whose unwavering support and guidance have been instrumental in nurturing our passion for quantity surveying. Their influence has laid a solid foundation for our contributions to the esteemed field of construction economics. Their belief in our potential and their endless encouragement have been pivotal in our journey.

Additionally, our acknowledgement extends to future generations of professionals in the construction industry. This work is dedicated to those who will carry forward the torch of construction economics development and study, innovation, sustainability, and societal advancement in the 21st century. It is their commitment to excellence and progress that will continue to shape and redefine the principles of construction economics in the years to come.

Our heartfelt appreciation goes to all who have played a part in our professional journey, inspiring us to strive for excellence in our contributions to this noble profession.

About the Authors

Andrew Ebekozien is a senior research associate in the Department of Construction Management and Quantity Surveying, University of Johannesburg, South Africa. He is also a lecturer in the Department of Quantity Surveying at Auchi Polytechnic, Nigeria.

Clinton Aigbavboa is the director of the NRF/DSI Research Chair in Sustainable Construction Management and Leadership in the Built Environment and of the CIDB Centre of Excellence at the University of Johannesburg, Johannesburg, South Africa.

1 The Construction Industry

1.1 NATURE AND SCOPE OF ACTIVITIES IN THE CONSTRUCTION INDUSTRY

Globally, the construction industry is significant because it is one of the most critical industries for the socio-economic development of developed and developing countries. The sector can influence most economic sectors (finance, industry, and commerce), contributing significantly to infrastructure advancement and gross domestic product (GDP). Also, it has been reported that the industry accounts for about 6% of the globe's GDP (World Economic Forum, 2018). Ofori (2012) found that the sector contributes between 3% and 10% of GDP and creates job opportunities for about 10% of a nation's employees. The construction sector is also known as the "built environment." This sector is pertinent to nations because it adds to personal satisfaction through its engineering and services provided to the end-users. Thus, it is a pertinent sector of every economy. Yet, the sector's impact is more pronounced in developing countries. This is because of the continuous demand for services to provide the basic infrastructure that will assist in measuring up to their developed countries. The Nigerian built environment sector is a remarkable, complex, and regularly divided industry. The construction industry is also known as the Building and Construction Sector in its sectoral classification of the economy. The built environment sector grasps a wide scope of inexactly incorporated associations that, on the builds, modify and fix a wide scope of the various building structure, structural designing, and heavy engineering work. Yet, the boundary between these regions is obscured. It incorporates planning, guidelines, planning, production, installation, and upkeep of structures and other structures. The sector has a duty regarding infrastructure advancement.

There are several attributes and structures of activities in the construction industry. These attributes are like the ones in the manufacturing and service industries. However, the former has physical goods. These goods are of mind-boggling size and price. This implies that the construction sector is a service industry. The structures and systems of activities in the sector interact to accomplish the desired construction goals. The sector is multifaceted and evidence of challenges, especially in developing countries. Nigeria's construction sector is confronted with project performance encumbrances (Unegbu *et al.*, 2022). This has negatively influenced the decline of the national GDP. Also, it has enhanced non-collaborative work practices in some construction projects (Olanrewaju *et al.*, 2020). Developing countries, Nigeria included, have been identified to be fertile ground for the construction industry (Akdag and Maqsood, 2019). But studies such as Nwachukwu and Nzotta (2010) and Ebekozien (2020) discovered that poor performance of construction projects (roads, bridges, buildings, dams, etc.) is regressive in many developing countries. This could

DOI: 10.1201/9781032661391-1

be traced to several factors, including corruption and lax utilisation of project management best practices. The industry accumulates little much capital compared to the manufacturing and processing industries. The built environment sector is an assembly sector, coupling on-site the finished goods of other sectors. Contract building plans and other relevant documents depict the Architect's originator goals. While the trained workers are guided by professional supervisors engaged in designing, constructing, and coupling various construction parts on site, division 45 of the Revised 2003 defined the construction industry by Standard Classification as follows:

i. General construction and demolition activities: organisation employed in building and civil engineering activities not specific to be ordered elsewhere.
ii. Construction and repair of building projects: organisation employed in developing, upgrading, and renovating private and non-private structures. This includes specialists employed in segments of development and maintenance work: for example, block-laying, the erection of steel, and concrete works, among others.
iii. Civil engineering: construction of pipelines, tunnels, runways, roads, railways, airports, etc.
iv. Installation of fixtures and fittings: organisation employed in the fixing of fixtures and fittings. This includes fittings, plumbing, and electrical fixtures.
v. Building completion: organisation employed in activities such as joinery, onsite tasks, plastering, painting, and decorating.
vi. Heavy engineering work: construction of power-generating plants, turbines, petrochemical factories, gas plants, refineries, farm tanks, cement plants, sugar plants, shipyards, aluminium plants, etc.

In developed nations, the construction sector is the largest employer of labour. In developing countries, such as Nigeria, the agricultural sector ought to be the largest employer of labour followed by the construction sector. This is not the case today in Nigeria. The agriculture industry is also known as the agriculture, forestry, and fishing industry. Therefore, the built environment is a critical factor or variable of steps forward in the drive for the economic development of nations, especially developing countries across the globe. Thus, construction activity is a global task and centres on infrastructural and industrial advancement (Asare *et al.*, 2022). In a study conducted by the World Bank in 2013, it was revealed that:

(i) Most Nigerians have inadequate access to basic public services such as pipe-borne water, roads, and electricity.
(ii) There is insufficient pipe-borne water, electricity, and roads in developed cities.

The construction industry cannot suggestively impact the demand for its outcome or regulate the supply. This indicates that several economic factors influence the degree of activity in the construction sector. Therefore, the need for a construction project is persuaded by the following variables:

(i) The built environment is susceptible to economic impacts, as witnessed by the downturn in the world economic meltdown in 2008. This is possibly one of the reasons the government regularly employs the construction sector as a mechanism to regulate the economy. This can be in the form of fluctuating interest rates to control the demand for residential buildings.

(ii) The public sector is the client/employer of about 50% of construction works. This makes it easy to cut public sector spending on construction works such as roads, hospitals, and school buildings.

(iii) Demand can come from a variety of sources. From the construction of one-bedroom apartments to mega projects such as the 2014 Brazil World Cup.

(iv) An optimistic construction market demands the accessibility of realistic cost credit.

(v) Also, state and federal governments can influence the demand for construction products. This can be done by allowing tax breaks for some construction activities.

(vi) The repair and maintenance section of the construction sector's output covers about 50%. This can be negatively affected in times of economic downturns.

The following variables categorise the supply side of the sector:

(i) Its unique structure: The sector's structure makes it hard to introduce new techniques and working practices to enhance output. This can be possible in larger firms with trained manpower, financial capability, and equipment, among others, to bring about change. Innovation in this sector could be expensive.

(ii) The trend of migration from traditional construction skills to mechanical or mobile construction with an emphasis on digital construction for the past 20 years. There is ongoing advancement in technological construction, for example, wooden windows and zinc coverings. Recent trends show a move towards an assembly process, for example, aluminium windows and roof covering.

(iii) The time interval between the reactions to supply to an expanded interest will almost consistently bring about a contortion of the market. For instance, increased residential houses react to expanded interest, activated by lower interest rates.

1.2 CHARACTERISTICS OF THE CONSTRUCTION INDUSTRY'S PRODUCTS

The construction sector has certain attributes. This is because of the sector's uniqueness (Mokhtariani *et al.*, 2017). Also, the product's physical nature and demands cannot be overemphasised. Some of the characteristics are as follows:

(i) No two construction projects are alike (one-off products). Even with prototype construction projects, project site attributes and location vary, making them unidentical.

(ii) Most times, the products of the construction sector are manufactured on the client's property, that is, the construction site (demand from the client's side).

(iii) Construction activities are conducted on the site where end-users will use the product.
(iv) Production is not conducted under controlled factory conditions most of the time during installation. Construction work is affected by the weather.
(v) Thus, the products are immobile.
(vi) Construction products take a long time to produce and last for many years.
(vii) Construction work requires huge financing.
(viii) The process that leads to production involves high risk.
(ix) Its procedures include a multifaceted combination of different materials, skills, and trades.
(x) The construction sector rendered service activities. Thus, it is more of a service industry than a manufacturing industry.
(xi) Globally, the construction sector includes a few large firms and a larger number of small construction organisations.

1.3 KEY PARTICIPANTS IN THE INDUSTRY

The construction sector can be sub-grouped into three major areas. This is building, civil, and heavy engineering work (process plant engineering). Each one complements the others. Figure 1.1 presents the major stakeholders in the construction industry. Figure 1.1 illustrates that major participants in the construction industry are divided into three major parts: the design team, client, and contractor, respectively.

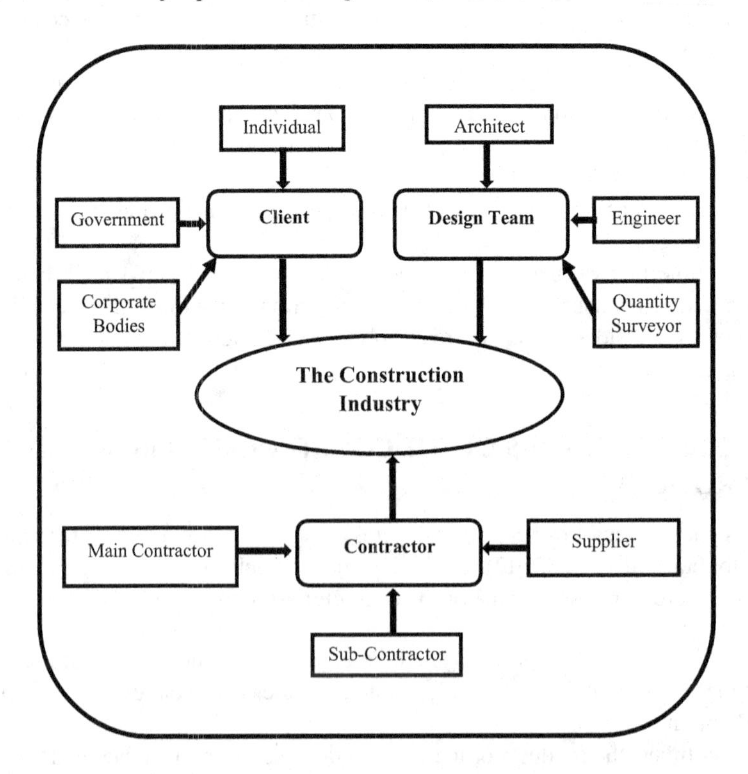

FIGURE 1.1 Major Participants in the Construction Industry

1.3.1 Design Team

1.3.1.1 Quantity Surveyor

"A Quantity Surveyor is a cost and procurement management expert concerned with financial probity and achieving value for money in the conceptualisation, planning, and execution of building, civil, and other heavy engineering projects" (Nigerian Institute of Quantity Surveyors, 1998). A Quantity Surveyor (QS) is the Cost Manager and Cost Engineer for construction projects. Among other functions are preparing bills of quantities (BOQs). This depends on the accuracy of the availability of full particulars of the work involved. Other contract documents prepared are the specification of materials and workmanship, or preambles, preliminary bills, and contract conditions. The QS' other duties can be summarised as follows:

(i) The QS is available to guide the employer and Architect on the proposed construction's cost-effectiveness. This includes matters arising from constructional aspects and the financial side of construction contracts.
(ii) The QS advises on the financial aspects of building, civil, and heavy engineering works during the pre-contract period, including feasibility studies.
(iii) The QS prepares an approximate construction cost estimate.
(iv) The QS prepares the BOQs, and where this is impossible because of inadequate information, schedules of prices or bills of approximate quantities are prepared.
(v) The QS decides on the pricing in BOQs or schedules of prices.
(vi) The QS prepares the documents necessary for procuring competitive tenders from construction contractors. One of the roles is to advise on the form of the contract and payment terms. This role is germane from the onset to avoid disputes in the contract in the future.
(vii) The QS evaluates, as the construction work is in progress, the work's value to date and makes recommendations to the Architect as to the payment on account to be made to the construction contractor. This could be based on milestones or periodic arrangements.
(viii) If necessary, the consultant QS reports to the client and copies other design team members on the financial impact of alterations (variations) ordered or proposed as deemed.
(ix) The QS conducts the technical checking of accounts presented by the construction contractor. These accounts are based on the construction contractor's costs.
(x) The QS is expert evidence on disputes.
(xi) The QS prepares and settles construction contractors' final accounts.

While the duties highlighted above may concern new buildings, there are other ancillary functions that the QS can perform on old or dilapidated buildings. Some of these are as follows:

(i) Priced BOQs preparation for the reinstatement of fire-damaged buildings
(ii) Assessment and settlement of claims for fire damages
(iii) Professional witness in cases of dispute

(iv) Preparation and pricing of schedules for dilapidated buildings
(v) Conciliation and settlement of claims for the property owners or tenants

Therefore, "quantity surveying is primarily concerned with the detailed calculations and assessment of quantities of materials and labour required for all construction activities such as building, civil and heavy engineering" (Nigerian Institute of Quantity Surveyors, 1998). The qualifications of a QS are no doubt extensive. The QS must be acquainted with building, civil, and heavy engineering construction, services, measuration, and allied subjects. As a cost adviser, technical accountant, and specialist on contractual matters, a QS is concerned with a building project from commencement to its ultimate financial settlement, which may be some considerable time after the project's completion.

1.3.1.2 Architect

The duties can be summarised as follows:

(i) The Architect should discuss with the client, collect briefs, and visit the site. The Architect then crystallises the brief into a factual foundation for the project.
(ii) The Architect ensures that the proposal complies with statutory requirements.
(iii) The Architect assembles the design team and heads the design team. This is not a right but is determined by many factors. For example, the scope of work and the client's discretion.
(iv) The Architect produces a sketch design and obtains the client's approval after checking with other consultants.
(v) The Architect checks with the QS to see that the cost matches the budget.
(vi) The Architect prepares tender/contract drawings and coordinates with other consultants.
(vii) The Architect supervises the construction as per agreement with the client. Certify work for payment at agreed intervals.
(viii) The Architect issues a practical completion certificate at the project completion stage.
(ix) Also, a certificate of making good is issued where defects/faults, if any, have been made good. This is usually after a six-month defect liability period.
(x) See that a final account is settled and the project is verified as completed.

1.3.1.3 Engineer

On any one project, there are usually many Engineers on board. On a building project, you would have a Structural Engineer, an Electrical Engineer, and a Mechanical Engineer. Briefly, their duties include the following:

(i) The Engineer works with other team members, including the Architect. The Engineer ensures that the proposal is adequate and can be served.
(ii) The Engineer produces relevant detailed calculations, schedules, and specifications to enhance the QS' BOQ preparation and cost control plans.
(iii) The Engineer ensures that the scheme satisfies statutory requirements, and carries out tests of soils, among other tests.

(iv) The Engineer supervises the project and modifies the scheme when necessary. For certain projects, like in full civil engineering works, the Civil Engineer becomes the main consultant and may not require the services of the Architect. In such cases, the outline of his duties includes obtaining a brief, preparing a programme, and supervising.

1.3.2 Client (Employer)

The client could be an individual, an organisation, or a government agency (federal, state, and local governments). Ideas for projects usually emanate from the client, except in speculative development.

(i) The client's duty starts with conceptualising a design brief. The brief sets out, possibly in clear terms, the components of the project. This includes what the project is aimed at. A clear brief is not always forthcoming, as some clients can never adequately articulate and outline their requirements.
(ii) The client needs to obtain the services of competent consultants to handle the project.
(iii) The client must have a reliable budget and predictable cash flow to comply with contract terms.
(iv) The client must assume his duties as per contract requirements. The client has to settle all payment certificates as they are due and take insurance as appropriate.

1.3.3 Contractor

Contractors range in size from a one-person outfit to large multi-national conglomerates. Also, they vary in size, capability, and steadfastness. Some contractors have sections that can handle all facets of construction, while others rely on sub-contractors' services for specialist jobs. Generally, the following are some of the expected duties of an engaged contractor:

(i) The contractor must accept, upon "winning" the contract, to handle it.
(ii) The contractor should conduct a site investigation for abnormal conditions.
(iii) The contractor gives the client and consultants more duplicates of the priced BOQs if required for ease of cost control.
(iv) The contractor prepares a pre-contract plan. The contractor also raises queries on drawings in cases of discrepancies.
(v) The contractor hosts regular site meetings.
(vi) The contractor provides site facilities for consultants and other site staff and support sub-contractors.
(vii) The contractor provides consultants with enough information to help them monitor progress.
(viii) The contractor takes insurance coverage as required.
(ix) The contractor should adhere to the contract conditions, rectify defects, and complete the work on schedule. The nominated sub-contractors and suppliers have obligations to the contractor. The duties include insurance, adherence to

the work programme, and ensuring that quality work is done and good materials are supplied. A discount on the contractor's certificates must be allowed. Suppliers should also present their accounts on time for settlement.

1.4 TYPES OF CONTRACTORS IN THE INDUSTRY

There is no broad ideal of the construction organisation, as the size is influenced by the idea of the project and the situations under which it should be completed. The firm's characteristics and the organisation's capability should be considered. For instance, maintenance and construction work in small houses require few workers. This would not be striking to big organisations. Therefore, construction organisations within the industry are classified according to the type of activities:

1.4.1 Large Contractors

Organisations in this category can embark on large building, civil, and heavy engineering projects nationwide and abroad. This is mostly within the continent. Opinions vary concerning how to group a large organisation, and the measurement scales differ from more than 300 workers to more than 1,200 workers.

1.4.2 Medium-Sized Contractors

They can be engaged in large construction projects within a 250-km range of their administrative locations. They engage in building and civil projects. They employ between 50 and 300 workers, and their operation is limited to the regional area.

1.4.3 Small Contractors

These established organisations operate within a close range of their administration locations and travel further afar only under extraordinary conditions. This size of organisation may experience an issue when the virtues of the founder are not inbuilt into the successor. In most cases, the successor is a very close relative, either the child or grandchild, responsible for the day-to-day running of the firm. They employ more than 50 workers, but fewer in most organisations in this category.

1.4.4 Specialist Sub-contractors

The specialist sub-contractors focus on one section of the construction or associated project. Thus, the sub-contractor is responsible for ensuring that the engaged expertise is trained with high-skill knowledge. This skill is utilised in operating the designated equipment and supervising operations. The outcome is exclusive performance and effectiveness with a decent quality-completed product. This type of sub-contractor can be comprehensively classified into three groups:

- Craft firms: There are many crafts or trades in the industry, for example, timber roofing, tiling, painting, plumbing, plastering, scaffolding, and glazing, among others. This group specialises in one of the building crafts.

- Constructional firms: Also, in the construction sector, works can be grouped into work sections, for example, concrete work, woodwork, groundwork, blockwork, steelwork, and piling, among others. This group specialises in one of the work sections.
- Service firms: This is designated for mechanical and electrical installation, for example, construction works associated with services, ventilating and air conditioning plants, lifts, escalators, and associated services, electrical installations, and heating, among others.

1.4.5 Labour-Only Contractor

These firms with small trade gangs work independently of any other firm but move from site to site and provide only labour. Labour-only contracting has been a reason for impressive debate over the years. The government and some scholars claim it prompts poor workmanship standards, tax avoidance, and sub-standard work conditions.

1.5 THE IMPORTANCE OF THE CONSTRUCTION INDUSTRY

The significance of the construction sector is derived from the following factors:

i. **Its size:** The construction sector in Nigeria is about 12% of the GDP. This is not a fixed figure, so the percentage fluctuates with the economy. In the last global recession, the construction sector was the first to enter the recession and one of the earliest to recover. The construction sector provides over 50% of the fixed capital investment in the economy (ships, vehicles, aircraft, plants, etc.). The construction industry experienced rapid growth in the 1970s. After the civil war, the recession in the mid-1980s had harsh repercussions, with output plunging. The effect was cushioned from 2000 to early 2014 but worsened (Olanipekun and Saka, 2019). At the beginning of the 21st century, the construction sector was optimistic, but this would be short-lived as the world faced an unprecedented recession. The Nigerian construction sector is the second-largest in Africa. It represents about 12% of the total output. The South African construction industry only exceeds it.
ii. **It provides investment goods predominantly:** It has been established that construction is an asset industry. The products from the construction sector are needed not for their own well-being but for the goods or services they can help make. This is particularly true for factory buildings, but it applies to many other building types.
iii. **The government is a large client**: In the early 1990s, public buildings accounted for over 50% of the construction industry's workload. Work is undertaken for various local, state, and federal governments, including ministries, departments, and agencies. They belong to this category.
iv. **Employment:** The construction sector engages approximately 7.5 million workers. Larger numbers of others are engaged indirectly by material and component manufacturers. A wider scope of secondary employment relies on a prosperous construction sector.

1.6 THE CONSTRUCTION INDUSTRY AS AN ECONOMIC REGULATOR

The Nigerian government remains a major client of the construction sector. Many scholars have argued that the construction sector is an economic regulator. The government may concede or drop projects for different reasons. For example, decreasing the public sector borrowing obligation in turn often makes a thump on impact. Public expenditure cuts could occasionally have a high construction effect yet are frequently joined by different measures (Abdullahi and Bala, 2018; Olanipekun and Saka, 2019). Whether these can be referred to as examples of regulation is debatable.

The government can mediate in the construction market in three different ways, through funds, enactment of regulation, and provision, as follows:

- Intervention in the market through the fund by grants, benefits, subsidies, and tax collection.
- Grants for industrial or commercial premises in regions/locations of high joblessness.
- Incentives for specific types of projects, for example, private residential housing.
- Taxation breaks against profits for the yearly maintenance of construction projects.
- Changing guidelines, for example, in town and country planning. This can generate opportunities for construction by permitting the extensive scope of activities in confined zones.
- Since the government is a noteworthy customer, it has an extensive degree of influence over construction activity through projects' development, repair, or maintenance.

In managing the economy, the government has four main objectives:

- The capability to pay its way aboard by balancing the payments
- A suitable level of engagement of resources, particularly manpower
- An increase in the number of goods and services produced and consumed leads to a higher standard of living
- The control of inflation

However, some economists have differing views about whether the government ever uses the industry as an economic regulator. This assertion is because of the break in the effectiveness of sudden stops and starts in its workload. Therefore, government measures to control the economy usually affect the construction industry. This is either directly or indirectly caused by the control of output or by restricting borrowing via high interest charges.

1.7 EFFECT OF THE INDUSTRY ON THE NATIONAL ECONOMY

The affluence of the Nigerian construction sector is firmly identified with the economy's condition. They both have a connection. In the extreme financial recession, the construction industry endured its most noticeably terrible post-common war during the structural adjustment programme in former President Babangida's government. The impacts of changes on the outcome, employment, incomes, or demand in the construction sector will probably have repercussions in other sectors of the economy (Olanipekun and Saka, 2019). Consequently, a decrease in construction activity will adversely affect many other industries. This sector is one of the major economic variables. One of the reasons is that the governments (local, state, and federal) are the main clients/employers as previously explained.

1.8 EFFECT OF GOVERNMENT ACTION ON THE INDUSTRY

The governments have an important role because they buy over half of the construction sector's output. Studies have shown that the fluctuating workload would hurt the construction sector, regarding the management and planning of the resources. However, government policies have caught up with the construction industry (Olanipekun and Saka, 2019). These policies either restrain or stimulate the economy. When government regulations restrain operations, the volume of public construction projects is reduced. The adverse effects of depression on the construction industry are as follows:

- Joblessness of construction workers.
- Smaller organisations are forced out of the industry.
- Large organisations are unwilling to invest in machinery with emerging innovations.
- Suppliers of construction materials are not willing to cooperate.
- Engagement of workers becomes harder.
- Increased costs and reduced efficiency because of a lack of continuity of work.

The chapter shall consider some of the repercussions of government policy:

(i) **The Government as a Client:** The Nigerian Government undertakes over 70% of the construction work. This is done by undertaking the construction or supporting the private developers with accessible funding to construct houses for the medium, low-medium, low-income earners. These government projects include electrification, roads, and building new and expanding towns. Government expenditure on roads, health-care centres, and schools, if curtailed, would have a serious impact on the construction sector. Also, diversion of construction projects budget for other tasks could be a threat to the industry.

(ii) **Public Control:** The government seems to persuade the location of industries or new works in certain ways:
- Construction subsidies and tax incentives
- Through grants or other incentives
- Location

(iii) Monetary Policy: The government utilises different controls, for example, bank rates, open market operation, and higher purchase restrictions, to adjust the degree of the interest rate. It will bring about higher prices for buildings in the long run. The outcome might discourage some housing developers and clients from investing in construction projects. The simple reason is inadequate profit to justify the huge investment. One of the possible outcomes is project postponement or abandonment because of difficulties in accessing funds from the banks.

(iv) **Taxation:** Taxation is a significant instrument in government fiscal policies. One of the reasons is that it is charged on capital and income. Therefore, a rise in capital tax on the value of the building or profits made from the business transaction of the building or land may bring about diminished interest. Also, increased property income may increase interest in the new building.

(v) **Fiscal Policy:** The government can persuade the degree of monetary activities by controlling expenditure. This would affect the construction industry because the government is the major client. However, during deflation and rising joblessness, increasing public works creates job opportunities to resuscitate the economy.

1.9 PROFESSIONAL AND REGULATORY BODIES

1.9.1 Professional Bodies

A typical professional body is an association limited by guarantee and registered under the Companies and Allied Matters Act. Its existence may be further recognised by an Act of parliament or decree that establishes the statutory organ of government meant to regulate the practice of the said profession (NIQS, 1998). For instance, Decree No. 31 of 1986, which established the Quantity Surveyors Registration Board of Nigeria (QSRBN), formally recognised the Nigerian Institute of Quantity Surveyors (NIQS) (NIQS, 1998). Globally, the professional bodies' requirements are high. So, a potential QS must be a university or polytechnic graduate. This is part of the academic requirements. The student would have been exposed to practical experience and professional education during the programme. There are limited opportunities for those who studied related quantity surveying disciplines who wish to change or qualify as corporate members.

Every intending corporate member will have to meet the NIQS requirements. This includes the Graduateship Examination (GDE), Test of Professional Competence, and Professional Competence Interview before they can be qualified as Professional Quantity Surveyors. *Note*: Only university graduates (B.Sc./B.Tech./M.Sc./M.Tech. in Quantity Surveying) are exempted from GE (NIQS, 1998). The Institute (NIQS) makes it compulsory for members to achieve a minimum of 20 units of Continuing Professional Development in any one year. This ensures that members are well-informed about new methodologies throughout their working lives, for example, the application of BESMM4 and Masterbill (Mb) software. This is made possible through seminars/conferences/workshops organised by the Institute and chapters (with the approval of the National Body). And at the end of the event, a press release is made, which forms the basis for the recommendations to policymakers and other stakeholders in the construction sector. Also, such a platform is utilised to sensitise members on current issues regarding the profession. Only elected fellows and corporate members of the Institute may show the grades of membership of the Institute by utilising the suitable designatory letters as suffixes as follows:

Past President	PPNIQS
Fellow Member	FNIQS
Corporate Member	MNIQS
Honorary Fellow	Hon.FNIQS

1.9.2 Regulatory Bodies

A typical regulatory body is an establishment of government aimed at controlling the practice of a profession. Its power is derived from the Act of Parliament or Decree establishing it. Control is achieved through the following functions of the body:

- (i) The minimum educational qualifications and experience requirements must be obtained by persons wishing to practice the profession.
- (ii) Registration of those who have met with (i) above is subject to payment of the prescribed fees.
- (iii) Maintenance of a nominal roll of all professionals duly registered to practice at any given time, subject to meeting the minimum requirements for continued enrolment and practice.
- (iv) Approval of the study syllabus and accreditation of tertiary educational institutions offering courses in the profession concerned.
- (v) Disciplinary action against erring professionals.

It is a matter of fact that the relevant professional bodies generated the causes of action leading to the establishment of many of the regulatory bodies in the industry. It is, therefore, not surprising that the membership of the management boards of the regulatory organs is always largely drawn from within the relevant professional bodies. All regulatory bodies in the industry, as organs of government, are located and coordinated under the Federal Ministry of Works, Housing, and Urban Development. The supervising ministry provides funding for the running and functioning of these bodies. However, getting this funding/grant from the supervising ministry is difficult. Table 1.1 shows the key professional institutes in the industry under their respective regulatory bodies.

1.10 NATIONAL, INTERNATIONAL, AND REGIONAL ORGANISATIONS

- i) The Royal Institution of Chartered Surveyors (RICS)
- ii) Association of Professional Bodies of Nigeria (APBN)
- iii) International Federation of Surveyors (IFS)
- iv) International Cost Engineering Council (ICEC)
- v) The Commonwealth Association of Surveying and Land Economy (CASLE)
- vi) The African Association of Quantity Surveyors (AAQS)
- vii) West African Construction Economics Association (WACEA)

TABLE 1.1
Professional Bodies with Their Functions and Respective Regulatory Organs in Nigeria

S/No	Professional Bodies	Main Responsibility	Designation of Corporate Members	Statutory Regulatory Body
1	Nigerian Institute of Estate Surveyors and Valuers (NIESV)	1) Estate management 2) Property valuation	ANIVS (member) FNIVS (fellow)	Estate Surveyors and Valuers Registration Board of Nigeria (ESVRBN)
2	Nigerian Institute of Architects (NIA)	1) Project design 2) Supervision and execution	Arch . . . MNIA (member) Arch. . . FNIA (fellow)	The Architects Registration Council of Nigeria (ARCON)
3	Nigerian Institute of Quantity Surveyors (NIQS)	1) Total cost management 2) Construction, economics, and planning	MNIQS (member) FNIQS (fellow)	Quantity Surveyors Registration Board of Nigeria (QSRBN)
4	Nigerian Institute of Builders (NIOB)	1) Project erection and supervision 2) Construction management	MNIOB (member) FNIOB (fellow)	Council of Registered Builders of Nigeria (CORBON)
5	Nigerian Society of Engineers (NSE)	1) Engineering designs (civil, structural, electrical, and mechanical) 2) Supervision of engineering component	Engr . . . MNSE (member) Engr. . . FNSE (fellow)	Council of Registered Engineering in Nigeria (COREN)
6	Nigerian Institute of Town Planners (NITP)	1) City and infrastructure designs 2) Control of development	MNITP (member) FNITP (fellow)	Town Planners Registration Council of Nigeria (TOPRECON)
7	Nigerian Institution of Surveyors (NIS)	Land and hydrographic surveys	ANIS (member) FNIS (fellow)	Surveyors Registration Council of Nigeria (SURCON)
8	Society of Construction Industry Arbitrators (SCIArb)	Arbitration	MSCIArb (member) FSCIArb (fellow)	

1.11 LATERAL COOPERATION BETWEEN BODIES IN THE INDUSTRY

Lateral cooperation between various professional bodies or construction federations and unions is a significant and constant activity within the industry. The cooperation is evidenced by the formation of various joint committees, which may be ad hoc or standing, depending on the purpose or task of the committee. Examples of such bodies are:

1.11.1 National Joint Industrial Council (NJIC)

This is a standing committee with representatives from the Federation of Building and Civil Engineering Contractors (FOBACEC) and the National Union of Construction and Civil Engineering Workers (NUCCEW). The council's function is to discuss and agree on several issues of mutual interest to the member parties, as the issues arise from time to time. One such issue is the "condition of engagement or contract of employment of workers in the industry." Also, government policies such as the national minimum wage, as they affect industries, are discussed and translated by the NJIC for ease of implementation by all concerned.

1.11.2 Building Contractors Joint Board of Nigeria (BCJBN)

The BCJBN, an ad-hoc board, comprised representatives of the Nigerian Institute of Quantity Surveyors, Nigerian Institute of Architects, Nigerian Society of Engineers, Federation of Building and Civil Engineering Contractors, and Nigerian Institute of Builders (Ebekozien, 2022). The board initially sets out to produce a Standard Form of Building Contract for the Nigerian construction industry, or the SFBN.

However, in its deliberations, the active involvement of the federal government through the Federal Ministry of Works, Housing, and Urban Development was sought, and this led to the inclusion of other professional bodies as observers and advisers. These include the Nigerian Institute of Estate Surveyors and Valuers, the Nigerian Association of Chambers of Commerce, Industry, Mines, and Agriculture, the Institute of Chartered Accountants of Nigeria, and the Association of Arbitrators of Nigeria as observers. Advisers were the Insurance Institute of Nigeria, the Nigerian Institute of Bankers, and the Nigerian Bar Association, with the Federal Ministry of Justice as legal advisers (Akindele and Adepoju, 2020). The involvement of representatives of almost all the relevant sectors of the economy as potential users and operators of the documents is quite significant.

1.12 TRADE UNIONISM

Historically, for the first time in employment law, legal rights of work for trade unions were empowered by the Employment Protection Act of 1975 in the United Kingdom. Before this Act of 1975, there were no legitimate rights. We had several agreements voluntarily with trade union activities. The Employment Protection Act of 1975 has made it simpler for employees to consult with their employers under one canopy (Nwoko, 2009; Adefolaju, 2013). This provides a platform for a better representative of both parties in matters of common interest, for example, welfare, safety, and health. The key responsibility of trade unions is to bargain regarding the employee's service conditions and remuneration rates. In Nigeria, it is called the NJIC agreement. Depending on government policies, this is normally revised every two years or as needed. This agreement is normally between construction and civil engineering. Employers Association of Nigeria and the National Union of Civil Engineering Construction Furniture and Wood Workers for Junior Employees in the Building and Civil Engineering Industry in Nigeria (Adeyemi and Aigbavboa, 2018). This is in addition to the high-rank officials in Nigeria's Building and Civil Engineering construction industry. The NJIC agreement for junior and senior employees was negotiated and agreed upon on Tuesday November 29, 2005. This was revised on December 5, 2007, and is now overdue for another review. Note that this agreement is the minimum standard. Many trade organisations are associated with the construction sector. Examples are the National Union of Civil Engineering Construction, Furniture, and Wood Workers, an amalgamation of many sub-unions for junior employees in the construction sector. Likewise, we have senior workers in the Building and Civil Engineering Construction Industry.

1.13 SUMMARY

This chapter introduced the readers to the structure of the construction industry and the inter-relationship of professionals in the built environment. The importance of industry follows this as an economic regulator and its impact on the national economy. The chapter gave an overview of the key functions of professionals and regulatory bodies in the construction industry. The chapter also highlighted lateral cooperation between bodies in industry and trade unionism. The next chapter gives an in-depth description of the formation of a quantity surveying firm and the types of practice. This includes the role of a QS in pre- and post-contract administration.

REFERENCES

Abdullahi, M. and Bala, K. 2018. Analysis of the causality links between the growth of the construction industry and the growth of the Nigerian economy. *Journal of Construction in Developing Countries*, *23*(1), pp. 103–113.

Adefolaju, T. 2013. Unions in Nigeria and the challenge of internal democracy. *Mediterranean Journal of Social Sciences*, *34*, pp. 12–18.

Adeyemi, B. and Aigbavboa, C. 2018. Assessment of factors affecting trade union in Nigerian construction industry. In *Proceedings of the international conference on industrial engineering and operations management*, Washington, DC, September 27–29.

Akdag, S.G. and Maqsood, U. 2019. A roadmap for BIM adoption and implementation in developing countries: The Pakistan case. *ARCHNET-IJAR: International Journal of Architectural Research*, *14*(1), pp. 112–132. https://doi.org/10.1108/ARCH-04-2019-0081.

Akindele, E.O. and Adepoju, A.O. 2020. Effect of project control practices on the performance of building construction firms in Lagos State, Nigeria. *African Journal of Science Policy and Innovation Management*, *1*(1), pp. 10–26.

Asare, E., Owusu-Manu, D.-G., Ayarkwa, J., Edwards, D.J. and Martek, I. 2022. Thematic literature review of working capital management in the construction industry: Trends and research opportunities. *Construction Innovation*. https://doi.org/10.1108/CI-09-2021-0177

Ebekozien, A. 2020. Corrupt acts in the Nigerian construction industry: Is the ruling party fighting corruption? *Journal of Contemporary African Studies*, *38*(3), pp. 348–365. https://doi.org/10.1080/02589001.2020.1758304.

Ebekozien, A. 2022. Construction companies' compliance to personal protective equipment on junior staff in Nigeria: Issues and solutions. *International Journal of Building Pathology and Adaptation*, *40*(4), 481–498.

Mokhtariani, M., Sebt, M.H. and Davoudpour, H. 2017. Characteristics of the construction industry from the marketing viewpoint: Challenges and solutions. *Civil Engineering Journal*, *3*(9), pp. 701–714.

Nigerian Institute of Quantity Surveyors (NIQS). 1998. *Directory of Members and Quantity Surveying Firms* (4th ed.). Lagos: NIQS.

Nwachukwu, C.C. and Nzotta, M.S. 2010. Quality factors indexes: A measure of project success constraints in a developing economy. *Interdisciplinary Journal of Contemporary Research in Business*, *2*(2), pp. 505–512.

Nwoko, K.C. 2009. Trade unionism and governance in Nigeria: A paradigm shift from labour activism to political opposition. *Information, Society and Justice Journal*, *2*(2), pp. 139–152.

Ofori, G. 2012. Developing the construction industry in Ghana: The case for a Central agency. A concept paper prepared for improving the construction industry in Ghana. *National University of Singapore*, *13*(1), pp. 3–18.

Olanipekun, A.O. and Saka, N. 2019. Response of the Nigerian construction sector to economic shocks. *Construction Economics and Building*, *19*(2), pp. 160–180.

Olanrewaju, O.I., Chileshe, N., Babarinde, S.A. and Sandanayake, M. 2020. Investigating the barriers to building information modeling (BIM) implementation within the Nigerian construction industry. *Engineering, Construction and Architectural Management*, *27*(10), pp. 2931–2958. https://doi.org/10.1108/ECAM-01-2020-0042.

Unegbu, O.C.H., Yawas, D.S. and Dan-Asabe, B. 2022. An investigation of the relationship between project performance measures and project management practices of construction projects for the construction industry in Nigeria. *Journal of King Saud University – Engineering Sciences*, *34*(4), pp. 240–249. https://doi.org/10.1016/j.jksues.2020.10.001.

World Economic Forum. 2018. *Shaping the Future of Construction—Future Scenarios and Implications for the Industry*. Geneva: World Economic Forum. Available at: http://www3.weforum.org/docs/Future_Scenarios_Implications_Industry_report_2018.pdf (accessed 20 March 2022).

2 Quantity Surveying Practice

2.1 FORMATION OF A QUANTITY SURVEYING FIRM

Most Quantity Surveying practices, especially the private firms, started in a small way. The organisation begins with the principal partner, a typist/clerk, and sometimes a messenger/cleaner. The PP would start with enough money to pay staff and office rent for the first six months. The PP might secure a job or not before setting up the office. However, he makes most of the connections for a job during the early months of operation. The connection may be through consultant designers (Architects or Engineers) or directly with potential clients. As the volume of work increases, office overhead may be increased by employing one or more Quantity Surveyors or assistants. Thus, the challenge of company structure arises depending on the volume of commission (consultancy) he has received. The staff strength may increase to six or more, including a secretary, who can be superfluous in job delivery. The principal's duty as the firm expands will be limited to the supervision of works of the staff, general administration of the firm, and ensuring that further commissions are secured. The boss is also involved in the editing/drafting of preliminaries. Other partners or staff Quantity Surveyor carries out the actual jobs of the firm (Ashworth *et al.*, 2013). A consultant firm may be grouped in terms of job commission (each construction project is headed by a Quantity Surveyor's partner). In some cases, a single job may be split into convenient sections, units, or groups, and each group handles a section of the unit (Adunfe *et al.*, 2022). For instance, a pre-contract commission may be grouped into:

i. Feasibility studies, preliminary estimating, and cost planning.
ii. Production of a tender document.
iii. Examination of tenders, reports, and negotiations.

2.1.1 Types of Practice

2.1.1.1 Independent Private Practice

Many private practices aim at giving proficient management of construction projects. They provide cost management services to the employer/client and design team members (Architect or Engineer) before and during construction. The quintessence of such a practice is that it acts firmly unbiased in the financial aspects of a construction contract. Being a client, designer, or contractor, it has no vested interest in maintaining and can act without fear or favour in coming to a settlement (Male, 1990; Musa *et al.*, 2010; Ashworth *et al.*, 2013; Peter *et al.*, 2019).

DOI: 10.1201/9781032661391-2

2.1.1.2 Consortium of Private Practice

In this practice, two or more organisations group together to offer services. This is common when there is a large construction project. The consortium usually includes several designers, including the Quantity Surveyor. In practice, the client is expected to pay the firm a unified fee or royalty. The system of payment strengthens the use of the term consortium. In most cases, a consortium of private practices is for a temporary purpose (Ashworth *et al.*, 2013; Peter *et al.*, 2019).

2.1.1.3 Integrated Private Practice

This is another way of saying that the consortium of private practices becomes a permanent arrangement by establishing a single practice. The firm covers all aspects of the design team, including the design and quantity surveying functions. This type of practice enhances teamwork among the members since they are literally in the building in most cases. It is necessary to state here that the depth of integration at the working level would determine the degree of achievement of the aim of this type of practice (Male, 1990; Ashworth *et al.*, 2013; Peter *et al.*, 2019).

2.1.1.4 Government Service Practice

The Quantity Surveyors Section/Unit is attached to the Architects or Engineers Department in some public services. Recent evidence shows that some government ministries/departments/agencies have modified the structure. The number of Quantity Surveyors engaged depends on the size of the establishment. There is an allegation that the public service Quantity Surveyor is not independent in some government agencies but lacks empirical evidence. The public Quantity Surveyor is a direct employee of an establishment. Every establishment is governed by rules and regulations. This may limit the Quantity Surveyor's freedom of action. This is possible in the daily routine of work execution and is not peculiar to the Quantity Surveyor as a profession in the built environment (Ashworth *et al.*, 2013).

2.1.1.5 Contractors Service Practice

Construction contractors tremendously require the Quantity Surveyor service. Many contracting firms engage Quantity Surveyors permanently. In some instances, a separate department/unit/section is created for the quantity surveying staff of the organisation. Quantity Surveyors in contracting services are naturally commercial in their viewpoint. They are out to make money for their employers. It should be noted that professional ethics should not be set aside, regardless of who your employer is as a professional Quantity Surveyor (Ashworth *et al.*, 2013).

2.1.1.6 Other Service Practice

Quantity Surveyors are found in other services that are not primarily in the five groups mentioned earlier. They are as follows:

Commercial Service Organisation: The Quantity Surveyors in the commercial service organisation are engaged by firms not directly dealing with the construction market. For example, banks and manufacturing companies engaged Quantity Surveyors as full-time staff for the supervision and contract administration of their construction projects. Others in this category are T & E Nig Ltd., First Aluminium, Tower Aluminiums, and WEMAPOD Estates, among others.

Student Service Practice: The Quantity Surveyors who are academicians belong to this group. This group of Quantity Surveyors ensures that the succession of professionals continues with innovations and inventions. They are the ones tasked with research on the different areas of the built environment, emphasising development sustainability. The professional bodies work in collaboration with Quantity Surveyors in academia to ensure that the profession expands into front-line global issues related to construction.

2.2 THE DESIGN PROCESS AND PROFESSIONALS INVOLVED

To avoid conflict at the pre-contract and post-contract stages of work, the Royal Institute of British Architects (RIBA) came up with an outline of procedure known as the RIBA Plan of Work (RIBA, 1887, 1963, 1972; Chappell and Dunn, 2015). This plan suggests a pattern of procedure for various professionals at both stages. There have been positive outcomes from many construction contracts subjected to this outline plan of work. The plan is outlined in Table 2.1.

TABLE 2.1
Outline Plan of Work

Stage	Aim	How to Accomplish the Aim	Parties Involved	Usual Terminology
A Inception	To prepare a general sketch of prerequisites and plan future action to accomplish the goal.	Set up the client's organisation for the briefing, consider prerequisites, and engage the Architect to accomplish the goal.	Employer and Architect	Briefing
B Feasibility	To provide the employer with an appraisal and suggestions to determine the form of the project. To ensure that it is feasible financially, technically, and functionally.	The design team studies user prerequisites, site conditions, planning design, and cost, among others.	Employer's representatives, Architects, Engineers, and Quantity Surveyors, according to the nature of the project, are involved at this stage.	

(Continued)

TABLE 2.1 (Continued)

Stage	Aim	How to Accomplish the Aim	Parties Involved	Usual Terminology
C Outline proposal	To determine the general approach to layout, design, and construction regarding the authoritative approval of the employer on the sketch proposal and accompanying reports.	The design team will develop a brief further. Also, conduct studies on user prerequisites, technical issues, planning, design, and costs to reach decisions.	All employer interests, Architects, Engineers, Quantity Surveyor, and specialists, as required, are involved at this stage.	Sketch plans
D Scheme design	To complete the brief and decide on the proposal, including planning to arrange, appearance, constructional approach, sketch specification, and cost. Also, to obtain all approvals.	The final development of the brief is expected at this stage. This includes the full design of the project by the Architect, the preliminary design by engineers, the preparation of the cost plan by the QS, and the full explanatory report. Submission of proposals for all approvals.	All employer interests, Architects, Engineers, Quantity Surveyors, and specialists. Also included are the statutory and other approving authorities.	
E Detail design	To obtain a final decision on every matter related to design, specification, construction, and cost.	Full design of every part and component of the building. Complete cost-checking of designs. After this stage, any further change or modification might influence the construction costs.	Architects, Quantity Surveyors, Engineers, and Specialists (contractors included If appointed).	Working Drawings
F Production information	To prepare production information and make final, detailed decisions to carry out work.	Preparation of final production information, that is, drawings, schedules, and specifications.	Architects, Engineers, and Specialists (contractors included if appointed)	
G Bill of Quantities	To prepare and complete all information and arrangements for obtaining a tender.	Preparation of bills of quantities and tender documents.	Architects, Engineers, and Quantity Surveyors (contractors included if appointed).	

(Continued)

TABLE 2.1 (Continued)

Stage	Aim	How to Accomplish the Aim	Parties Involved	Usual Terminology
H Tender action	To evaluate potential contractors and specialists. Also to obtain and appraise tenders.	Recommend contractors and specialists for the work.	Architects, Quantity Surveyors, Engineers, Contractors, and Clients.	
J Project planning	To provide planning of the site and methodology for construction activities.	Issuing information to the contractor. Arranging the handover of the site. Establish the site's layout. Also, prepare the programme of work.	Contractor, Sub-Contractor, Employer, and Architect.	Site operations
K Operations on site	To provide a procedure for construction activities.	Establish the methodology of construction, whether manually or mechanically.	Architects, Engineers, contractors, Sub-contractors, Quantity Surveyors, and Employers.	
L Completion	To ensure the construction project is achieved within the time and costtargets.	Administration of the building contract to practical completion.	Architects, Engineers, contractor, Quantity Surveyors, and Employers.	
M Feedback (post practical completion]	To analyse the management, construction, and performance of the project. Also, a review of project performance in use.	Administration of the building contract and making final inspections. This is key for future construction administration regarding the approximate estimation of similar projects.	Architects, Engineers, Quantity Surveyors, Contractors, and Employers.	

2.2.1 Stage A: Inception

It is necessary to identify the employer's request at an early stage. It is necessary to align this request with the financial, technological, and legislative. The Architect may start his work once the employer puts his request in a clear picture. The Architect may experience issue(s) receiving information from some employers. Seeley

(1997), RIBA (1997), and Hughes (2003) identified certain fundamental details that are common to construction projects as follows:

i. The planned construction project's nature, scope, and purpose
ii. The period and monetary limits compared to the construction projects
iii. Information connecting to the project site possession and other related issues
iv. Present state regarding any development application
v. Other individuals on the design team to be engaged

2.2.2 Stage B: Feasibility

The primary goal of this phase is to evaluate the proposed project's financial, functional, and technical aspects. The employer is advised regarding the feasibility of the proposed project. This is kicked off by engaging the design team in a meeting to discuss the employer's requirements in detail. The outcome of the meeting will produce a document on which the Architect assembles. The document can be used to appraise the feasibility of the proposed project. At this stage, there is clear information passed to the employer. The information will aid the employer in deciding whether to proceed with the project. The document would be presented in illustrative figures, such as plans and elevations, including cost estimates (Seeley, 1997; RIBA, 1997; Hughes, 2003).

2.2.3 Stage C: Outline Proposals

During this stage, the employer should have given the go-ahead to continue. The alternative scheme is arranged and contrasted to decide the general way to outline, plan, and construct. This includes close coordination between the employer and the design team. Each member of the design team has a significant role to play. The Architect will be worried about useful and stylish aspects. While the Structural Engineer centres around the distinctive structural forms. The Quantity Surveyor will keep close contact with the different specialists on the design team. This guarantees the substitute suggestions' cost implications as they are prepared and checked (Seeley, 1997; RIBA, 1997; Hughes, 2003).

2.2.4 Stage D: Scheme Design

At this stage, the design team critically examines the broader issues of appearance, the technique of construction, sketch specification, and the proposed cost plan. The proposed spread of cost over the elements or components of the building is considered. The Services and Structural Engineers will build up their work satisfactorily for the Architect to apply for full planning permission (Seeley, 1997). At this stage, all pertinent design information is assembled as part of the outcomes of the design team meetings. Then, a detailed document on the design and proposed cost implications can be presented to the client. The client's endorsement is required before the plan is further developed (Seeley, 1997; RIBA, 1997). The brief cannot be adjusted without the likelihood of incurring additional expenses after this stage.

2.2.5 Stage E: Detail Design

During the detailed design stage, the design will be developed further. This will include details to enhance the final decisions on the building components. The Quantity Surveyor will be kept hectically connected to prepare estimates of cost for the proposed construction project options (Seeley, 1997). This will be followed by continued cost checking on the cost plan to guarantee that the client's budget is not exceeded. At this stage, the client will be provided with a detailed design document, including the reviewed cost plan (RIBA, 1997). This is a critical stage. Once the client acknowledges the detailed design, he subscribes to the proposals. Any modification that may be required afterward will result in failed work and may include an extra cost to the client.

2.2.6 Stage F: Production Information

At this stage, the Engineers, Architects, and other specialists will continue preparing the final contract drawings and schedules. This will include supporting specification notes. Detailed information is passed to the QS, including building components, for example, doors, windows, ironmongery, and manholes, in the form of schedules. The production documents include the Architect's drawings, the Structural Engineer's drawings, and Service's drawings. It is standard to decide on the preliminary tendering procedures and prepare a list of possible main construction contractors. Other documents prepared are invitations to tender, information for tender documents, amendments to the JCT Form of Contract or FMW Condition of Contract, procedures for selecting subcontractors and suppliers, and maybe preparation of advance orders to sub-contractors and suppliers. Some of these tasks will require further meetings with the client (Seeley, 1997; RIBA, 1997; Hughes, 2003).

2.2.7 Stage G: Bills of Quantities

All relevant tendering documents, including the bill of quantities, are finalized during this phase. The QS is accountable for continually checking during the measurement of the work. The QS prepares a query list for the Architects and Engineers covering the omission of information, discrepancies, and ambiguities (Seeley, 1997). In response to the query list, the Architect and Engineer can clarify the points raised and amend the documentation before the tender stage. The Quantity Surveyor verifies prime cost and provisional sum sufficiency, respectively. This is to ensure that all important information is included. The bill of quantities offers valuable pricing and consultation on variations for interim certificates. It is part of the tendering documents and a good platform for cost planning, including the location and identification of the work. The bill of Quantities encompasses preliminaries, descriptions of materials and workmanship, and the measured work. While the preliminaries describe the scope and nature of the work, they also include details of the contract conditions, a list of drawings, and instructions to the contractor (Hughes, 2003).

2.2.8 Stage H: Tender Action

The QS, as the chief cost management expert in construction projects, analyses the tenders received (RIBA, 1997). This is followed by updating information for the client and consultants. Tender procedures should follow the prescribed major procedure in the Public Procurement Act 2007 as amended. The core ingredients of the rules that the "Due Process" seeks to ensure compliance with are as follows:

i. Appropriation
ii. Advertisement
iii. Pre-qualification procedure and criteria
iv. Invitation to tender
v. Opening of tender
vi. The bid evaluation process
vii. Determination of the winning bid

2.2.9 Appropriation

This absence in the old procurement system led to many construction projects being abandoned across Nigeria. Before 2007, Nigeria lacked a comprehensive legal framework or legislation on public procurement. For any construction project to occur, it should have been appropriated for in the budget. This policy encountered several encumbrances. Therefore, the Budget Monitoring and Price Intelligence Unit (BMPIU) was established in 2001 to tackle shortcomings associated with the policy. Treasury Circulars guided BMPIU operations based on the 1958 Act authorising the Accountant General of the Federation to issue guidelines on Public Expenditure. To institutionalise the operations of the BMPIU, the Public Procurement Bill was passed into law by the National Assembly on May 31, 2007. While the late President Yar'Adua GCFR signed into law the Public Procurement Bill on June 4, 2007. The Public Procurement Act put Nigeria in the league of countries with legislation on expending public funds. However, some scholars are of the view that the Act has enhanced corruption as against mitigating corrupt practices. This concerns the institutional framework and possibly lax enforcement of the Act.

2.2.10 Advertisement

There are two categories of advertisements based on the value of the contract.

i. Contracts below N10 million: The advert must be displayed on the public board and conspicuous locations within the ministry, parastatal, or agency (procuring entity).
ii. Contracts above N10 million: This need calls for pre-qualification of bidding contractors in at least two national newspapers or the *Federal Tender Journal*.

2.2.11 Pre-Qualification Process

The standard format set up by the Due Process contains the following:

i. The identity of the procuring entity is its name and address.
ii. A brief background description of the project, scope, commencement time, expected completion time, etc.
iii. Synopsis of the essential pre-qualification entering.
iv. Location and deadline for the return of filled-out applications for pre-qualification.
v. The time and date of the obtainability of the pre-qualification documents.

2.2.12 Criteria for the Pre-Qualification

The criteria comprise scores totalling 100% based on the following items:

a. Prove business name registration or incorporation 0% (Corporate Affairs Commission).
b. Registration with the Federal Ministry of Works in relevant category 0%.
c. The organisation audited its account for the past three years at 0%.
d. Prove tax clearance certificate for the past three years at 0%.
e. Prove the financial capability and banking support equal 15%.
f. Experiences/technical qualification (competence) and experience of key personnel = 25%.
g. A similar project was executed, and evidence of knowledge of the industry = 20%.
h. Annual turnover = 5%.
i. VAT registration and evidence of past VAT remittances = 5%.
j. Bonus = 3%. The components of the bonus are as follows:
    - Compliance evidence with the federal government and local content policy for building indigenous capacity = 5%.
    - Community social responsibility evidence.

Note: The variables with a zero score are measured as responsive and fundamental to the construction contractors, and the non-appearance of any of them will inevitably exclude an applicant for pre-qualification.

2.2.13 Pre-Qualification Benchmark

A benchmark of 70% and above is the best practice. A construction contractor is deemed qualified and competent if he scores above 70%. This is one of the guaranteed tests that inspires assurance in the bid result. The list of all the respondents who scored within and above 70% is drawn. This list is called the "list of pre-qualified bidders" or "list of competent bidders." All those respondents who fell below 70 are considered incompetent.

2.2.14 Invitation to Tender

The tender documents containing satisfactory information to enable contest among the bidders will be issued to the successful bidders who have decided to continue the bid process. From the date of release of the bid document and submission, the maximum time allowed is six weeks. Only in a few cases is the submission date extended, with the reason(s) made known to the bidders.

2.2.15 Opening of Tender

This happens directly after the closure of bidding/tendering. The opening of the tender takes place after six weeks of the issuance of tender documents. The bidders, representatives, and other stakeholders are invited during the opening of tenders.

2.2.16 Evaluation of Tender

The tender evaluation committee constituted by the procuring entity (ministry/parastatals/agency) shall appraise and prepare a report with a recommendation for award. The recommendation is submitted to the accounting authority (permanent secretary or chief executive officer) within the procuring entity.

2.2.17 Determination of the Winner

This is based on the recommendations of global best practices. The lowest evaluated tender cost by the requisite technical competence backed with sound financial capability is adjudged the right winner. The bid submitted by the winner is termed the "Best Response Bid."

2.2.18 Validation

After determining the winner, the procuring entity will submit a report of its handling procurement process to the BMPIU Compliance Review Unit for certification. If the report sails through, a validating certification is issued upon which final approval for the contract award is secured.

2.2.19 Stage J: Project Planning

Once the client and letter have approved the construction project of the award released to the contractor, preparing a programme for the operations will be necessary. This should be from the start to commissioning. Several useful techniques exist for programming, for example, networking and bar charts. Others are linear programming, transportation and assignment problems, Queueing theory, inventory control, simulation, and scheduling, among others. The suitable technique will allow the construction project to be appropriately controlled regarding time-related factors. Both the design team and construction contractor must work out the appropriate planning techniques. Note, planning the site layout is primarily the duty of the construction contractor, but the design team could advise if necessary.

2.2.20 Stage K: Operation on Site

During this phase, the construction contractor oversees activities at various stages of the construction project. The Quantity Surveyor measures the work done, either milestone or monthly, as agreed. This is known as an interim valuation and is forwarded to the Architect for onward release of the interim certificate to the construction contractor. The Architect has a responsibility to copy the Quantity Surveyor. This authority mandates the client to pay the construction contractor within the time frame stipulated in the contract conditions.

2.2.21 Stage L: Practical Completion

FMW Conditions of Contract refer to the practical completion of the works as a date to be determined by the Architect when he is required to issue a certificate clearly stating that the contract has reached this specific stage. In determining the date of practical completion of the works, the Architect should be completely satisfied with the answers to the following questions:

i. Has the work been carried out in accordance with the contract document and the Architect's instructions?
ii. Is the completed structure in a suitable state to be taken by the client for its full and proper use?

Once the certificate of practical completion is issued, the following take effect as agreed in the contract:

1. The construction contractor becomes eligible to receive payment of one moiety of the total of the retention fund (FMW Clause 30 (4) (b).
2. The defect liability period begins (FMW Clause 15).
3. The construction contractor is excused from the obligation to secure insurance for the works in accordance with FMW Clause 20A.
4. The period of final measurement begins (Clause 30(5).
5. Matters for arbitration depend on issuing the certificate of practical completion. This can be pursued.

Note: The Architect may be prepared to issue the certificate of practical completion but should detail to the construction contractor those incomplete items. This may apply to defective work, which the contractor must remedy immediately. The Architect's certificate should include a schedule of uncompleted work and defective items, if any. However, when the Architect is satisfied that the defective work has been made good, the Architect must release a certificate to the contractor. The Architect can consider the completion of the contract. Completion is defined as the final certificate. Clause 30 requires that the final certificate be released within three months of whichever of the following is the latest.

1. The end of the defect's liability period or
2. The completion of the making good of defects or
3. The receipt by the Architect/Quantity Surveyor of the documents from the contractor relating to the accounts of nominated sub-contractors and nominated suppliers (Seeley, 1997).

The Quantity Surveyor needs to forward the final account to the Architect. This will enable him (the Architect) to issue his final certificate. The approach and format the final account should take are documented in the Conditions of Contract. The Quantity Surveyor is responsible for preparing the details of the final account in the manner best suited for each project. However, the following are components of the adjustment of the contract sum in the final account:

1. Variations
2. Re-measurement of provisional quantities in the bill of quantities
3. Nominated Sub-contractor's account
4. Nominated supplier's account
5. Loss and expense caused by the disturbance of the regular progress of work (Clause 24)
6. Fluctuations in labour rates and material prices (Clause 31) (if applicable)

In preparing the final account, the Quantity Surveyor should provide all reasonable facilities for the contractor to be physically present when measurements and details are taken or recorded (Seeley, 1997). The Quantity Surveyor should forward a copy of the final account prepared to the contractor and the Architect. Once the final account has been prepared by the Quantity Surveyor and the Architect is satisfied with all other points previously referred to, he can issue his final certificate. This certificate releases the second moiety of the retention fund to the construction contractor. Hence, the final certificate amount is the final account's gross amount, less the amount of all previous interim payments. The final account stage of a contract is when the Quantity Surveyor determines the final cost of a construction project. This is based on the following documents:

- The form of the contract
- Original priced bill of quantities
- Variations
- Drawings
- Agreed contractor's claims

While the final account is composed of the following:

- Statement of the Final Account
- Final Account Summary
- Adjustment of Provisional Sums
- Adjustment of Prime Cost Sums if used
- Adjustment of Provisional Items
- Adjustment of the Variation Account

- Adjustment for Fluctuations
- Adjustment for Contractor's Claims (if applicable)

2.3 REPORT ON TENDERS

Report 1

EHI and OSE PARTNERS
(Project Managers and Registered Quantity Surveyors)
30, Igbe Road, Benin City, Nigeria.
08027000001, 054–257567

Our Ref: DD/A/00/010 February 12, 2023
Your Ref:

The Chairman,
Tender Board Committee,
XYZ Agency, Nigeria.

Dear Sir,

Proposed Engineering Building for XYZ Agency, Nigeria

About seven (7) weeks ago, XYZ agency announced an invitation to interested and qualified bidders on the pages of the Print Media; "Punch," November 29, 2022, and "The Federal Tenders Journal, November 29, 2022, to select tenders with adequate capabilities and structures considered relevant to deliver the project.

However, out of sixteen (16) bidders who completed and submitted the prequalification documents for the project, the Tender Board Committee of the XYZ Agency recommended Eight bidders. The bidders were eventually approved and selected by the Resident Due Process Team (RDPT) of the XYZ Agency on January 17, 2023.

This exercise witnessed five (5) tenders, which collected the documents relating to Building/Civil works. In line with instruction(s), each completed tender documents was submitted on or before 12.00 noon on December 29, 2022. According to records, one of the tenderers protested regarding time constraints.

The completed bidding documents relating to the project were publicly opened at 3 p.m. prompt in the presence of the bidder and/or representatives, an Education Trust Fund representative, members of the RDPT, including Messrs. Ehi and Ose Partners (Consultant Quantity Surveyors), and others in attendance.

Five (5) tenders were received after endorsements by the Chairman of the Governing Council and the director, who is also the Chairman of the Tender Board Committee. The documents were handed over to Messrs. Ehi and Ose Partners for further action.

The official records of the tenders for Building/Civil works are thus:

Proposed Engineering Building Breakdown of Tenders Received

No	Name of Tenderer	Tender Sum	Preliminaries	Total of Builder's Work	% of Builders Work from Tender Figure	Completed Period
1	Sam Ltd.	28,045,005.45	1,100,000	24,399,529	87.00	30 weeks
2	Peter Ltd.	26,135,318.32	275,000	25,136,558.70	96.18	20 weeks
3	Paul Ltd.	26,629,194.90	968,000	23,183,138	87.06	20 weeks
4	Monday Ltd.	27,108,739.35	1,100,000	23,529,847	86.80	35 weeks
5	Ikorodu Ltd.	20,877,595.20	687,500	17,985,924	86.15	12 weeks

Analysis and Comments on Tenders on Tenderer:

01 Tenderer: Messrs. Mr. Sam Ltd.

Analysis

Tender Figure:	₦28,045,005.45
Corrected tender figure	₦28,220,000.45
Arithmetical errors:	₦175,005.00
Percentages errors:	1.0%
Project quantity surveying estimate:	₦26,816,428.49
% below/above PQS estimate:	
(Using corrected tender figure)	5.5% (above)
Completion period:	30 weeks
Project quantity surveyor recommended award range + 5% of PQS estimate, i.e.:	₦25,475,607.07 to ₦28,157,249.91

Comments

This tender appears not to have been professionally priced. It showcases errors revolving around multiplications and extensions. It further parades unrealistic low/high rates for key elements such as concrete (1:2:4)—N16,000/m³, flush door—N45,000; steel reinforcement N290,000/tonne; N5,000/m² of roof covering, thereby exhibiting a non-coherent pricing policy. These ingredients are dangerous to post-contract administration. In addition, the corrected tender figure of N28,220,00.45 is 5.5% above the PQS estimate and falls outside the + 5% recommended award range. Also, the completion period of 30 weeks is on the high side.

02 Tenderer: Messrs. MR. Peter Ltd.

Analysis

Tender figure:	₦26,135,418.32K
Corrected tender figure:	₦26,295,285.62K
Arithmetical errors:	₦159,867.30K

(Continued)

(Continued)

Percentages errors:	0.7%
Project quantity surveying estimate:	₦26,816,428.49
% below/above PQS estimate:	
(Using corrected tender figure)	2% (below)
Completion period:	20 weeks
PQS recommended award range + 5% of PQS estimate, i.e.	₦25, 475,607.07 to ₦28,157,249.91

Comments

This tender is associated with several errors around additions, multiplications, and transfers. The tenderer pricing policy is inconsistent and unrealistic to a great extent. Thus, most key items are lowly priced and sometimes attract different rates for the same element. Concrete work N25,000/m3 in the substructure is low against N30,000/m3 and N33,000/m3 in the superstructure, formwork to columns N100/m2. These are never desirable, especially deeming post-contract activities. The corrected tender figure is 2% below the PQS estimate and falls within the + 5% recommended award range. The completion period of 20 weeks is on the fair side.

03 Tenderer: Messrs. Mr. Paul Ltd.

Analysis

Tender figure:	₦26,629,194.90K
Corrected tender figure:	₦26, 129,022.15
Arithmetical errors:	₦500,178.75
Percentages errors:	2%
Project quantity surveying estimate:	₦26,816,428.49
% below/above PQS estimate:	
(Using corrected tender figure)	2.6% (below)
Completion period:	20 weeks
PQS recommended award range + 5% of PQS estimate, i.e.:	₦25,475,607.07 to ₦28, 157,249.91

Comments

This tender accommodates moderate, evenly distributed rates. However, errors relating to extension addition and transfer resulted in the downward movement of the tender figure to N26, 129,022.15. The corrected tender figure of N26,129,022.15 is 2.6% below the project Quantity Surveyors (PQS) estimated and falls within the + 5% recommended award range. The completion period of 20 weeks is fair.

04 Tenderer: Messrs. Mr. Monday Ltd.

Analysis

Tender figure:	₦27,108,739.35
Corrected tender figure	₦27,108,739.35
Arithmetical errors:	Nil

(Continued)

(Continued)

Percentages errors:	Nil
Project quantity surveying estimate:	₦26,816,428.49
% below/above PQS estimate:	
(Using corrected tender figure)	1% (below)
Completion period:	35 weeks
PQS recommended award range + 5% of PQS estimate, i.e.:	₦25,475,607.07 to ₦28,157,249.91

Comments

This tender is noted for non-adherence to the stated requirement(s). Pencil instead of ink was used in pricing. The tender exhibited a high degree of desperation towards winning the contract with outrageously low rates on virtually all the items in the bills of quantities. Reinforced concrete (1:2:4)—₦14,000/M^3, reinforcement ₦85,000/ton, 225mm blockwork ₦4,500/m^2. The tender did not accommodate any form of error. The corrected tender figure of ₦27,108,739.35 is 1% above the PQS estimated and falls within the + 5% recommended award range. Still, the 35 weeks stated for the completion period is on the high side, and the tender may lack the capability to deliver the project.

05 Tenderer: Messrs. Mr. Ikorodu Ltd.

Analysis

Tender figure:	₦20,877,595.20
Corrected tender figure:	₦20,877,595.20
Arithmetical errors:	Nil
Percentages errors:	Nil
Project quantity surveying estimate:	₦26,816,428.49
% below/above PQS estimate:	
(Using corrected tender figure)	22% (below)
Completion period:	12 weeks
PQS recommended award range + 5% of PQS estimate, i.e.:	₦25,475,607.07 to ₦28, 157,249.91

Comments

This tender is soaked with several alterations without endorsement. The rates are evenly spread, consistent, and devoid of arithmetic error. However, most key items are lowly priced. This concrete (1:2:4) is rated ₦11,000/m^3 reinforcement—₦78,000/tonnes. Although the tender did not accommodate any error, the corrected tender figure of ₦20,877,595.20 is 22% below the PQS estimate and falls outside the + 5% recommended award range. The tender cannot deliver the project. The 12 weeks stated for the completion period are on the low side.

General Remarks/Recommendations

The tender, as corrected and analysed, cannot be associated with competitiveness as the difference between the highest and lowest tender is ₦7,167,410.25 (seven million one hundred and sixty-seven thousand, four hundred and ten naira and twenty-five kobo). Furthermore, tender figures associated with 3 (three tenderers) fall within the recommended award range and can be capable of delivering the project with the client paying more profit to the tenderers in ascending order of the affected figures/least evaluated tenders.

i. Messrs. Paul Ltd. - ₦26,129,022.15k
ii. Messrs. Peter Ltd. - ₦26,295,285.62k
iii. Messrs. Monday Ltd. - ₦27,108,739.35k

Arising from above, and in line with the "Due Process" Policy, we are inclined to recommend that Messrs. Paul Ltd. be awarded the project at the sum of ₦26,129,022.15k (Twenty-six million, one hundred and twenty-nine thousand, twenty-two naira, and fifteen kobo) at a completion period of 20 weeks. However, Messrs. Peter Ltd. should be offered the project should Messrs. Paul Ltd. decline at the same corrected figure of ₦26,129,022.15k (twenty-six million, one hundred and twenty-nine thousand, twenty-two naira, and fifteen kobo).

Thank you.

Yours faithfully,
For: Ehi and Ose Partners

A. Asemota (mniqs)
(Principal Partner)

Report 2

BROWN and CO ASSOCIATES
(Project Managers and Registered Quantity Surveyors)
30, Mission Road, Benin City
08027000001, 054–257567

Our Ref: DD/A/00/200 March 26, 2023
Your Ref:

The Chairman,
Tender Board Committee,
ABC Institution, Emuhi, Ekpoma,
Edo State, Nigeria.

Dear Sir,

Tender Report on Construction of Sports Pavilion and Upgrading of Sports Facilities for ABC Institution, Nigeria

Introduction

Following the tender advertisement in the Nation, Nigerian Tribune Newspapers, and Federal Tender Journal on January 20, 2023, eight contractors submitted their prequalification documents. Four contractors were prequalified and shortlisted to tender for the project.

The contractors shortlisted by ABC Institution submitted their tenders based on architectural/engineering drawings and bills of quantities prepared for the project. The contractors submitted their tenders before the close of the tender on February 15, 2023.

Tenders were opened at 1 p.m. prompt on March 3, 2020, in the ABC Institution Boardroom. The representatives of the Centre for Organisation and Professional Ethics (COPE-Africa), Servicom, Tenderers, TETFUND, and consultants were all present.

The tender figures were recorded as follows:

Tender Result

No	Name of Tenderer	Tender Sum	Completed Period
1	Sunday Ltd.	164,440,626.00	24 weeks
2	Clifford Ltd.	150,363,444.00	24 weeks
3	Musa Ltd.	163,061,965.50	26 weeks
4	Mar Ltd.	162,919,323.00	28 weeks
5	Consultant QS	149,750,286.00	28 weeks

The tenders handed to us were checked arithmetically, and the corrected figures are stated as follows:

Tender Result

Name of Tenderer	Tender Sum	Corrected Tender Sum	Error	% Error
Sunday Ltd.	164,440,626.00	164,414,166:00	+ 26,460:00	+ 0.016
Clifford Ltd.	150,363,444.00	151,386,805:50	-1,023,361:50	- 0.68
Musa Ltd.	163,061,965.50	159,818,200:50	+3,243,765:00	+ 1.989
Mar Ltd.	162,919,323.00	161,726,943:00	+1,192,380:00	+ 0.73

Analysis of Individual Tender

Sunday Ltd.

The contractor submitted the highest tender in ₦164,440,626:00 with a completion period of 24 weeks. Their tender is 9.81% above the Consultant Quantity Surveyor's estimate. A computation error of ₦26,460:00 brings the corrected tender figure to ₦164,414,166:00. Their tender is generally well priced except for fencing, laterite earth filling, and seeding; they were overpriced and therefore inconsistent with prevailing market prices. These accounted for the high tender submitted by the tenderer. We are not recommending them for further consideration on the grounds of high tender.

Clifford Ltd.

The tenderer submitted the most competitive tender in the sum of ₦150,363,444.00. This is 0.409% higher than the consultant's estimate. A computation error of ₦1,023,361.50 brings the corrected tender figure to ₦151,386,805.50. The tender is generally well priced, consistent with market prices. The rates which compare favourably with those of the consultants will ensure realistic cash flow during the currency of the contract. Their completion period of 24 weeks is optimistic and can be achieved. We will shortlist them for further consideration.

Musa Ltd.

The firm submitted a tender in the sum of ₦163,061,965.50. A computation error of ₦3,243,765.00 brings the corrected tender figure to ₦159,818,200.50. This is 6.723% above the Consultant Quantity Surveyor's estimate. This is considered too high to be shortlisted.

Mar Ltd.

This contractor submitted a tender figure of ₦162,919,322.00 with a completion period of 28 weeks. A computation error of ₦1,192,380.00 brings the corrected tender figure to ₦161,726,943.00. This is 7.998% above the Consultant Quantity Surveyor's estimate. Their tender is generally well prepared but with high rates. This accounted for their high tender, as shown in the detailed breakdown.

General Comments

Messrs. Sunday Ltd., Musa Ltd., and Mar Ltd. are good contractors from the above tender analysis. Their tenders are generally well prepared but are quite high considering the confident limit of ± 5% of the Consultant Quantity Surveyor's estimate. They are not shortlisted for further consideration on the grounds of a high tender. Messrs. Clifford Ltd. is shortlisted for further consideration. They are among the good contractors and have a cost advantage of 9.36%, 8.44%, and 8.35% over Messrs. Sunday Ltd., Musa Ltd., and Mar Ltd.

Conclusion

Based on the various tenderers' comments, Messrs. Clifford Ltd. is presented to ABC Institution for final consideration at the Consultant Quantity Surveyor's estimate of ₦149,750,286.00 (one hundred and forty-nine million, seven hundred and fifty thousand, two hundred and eighty-six naira) with a completion period of 24 weeks. We hope the report and recommendation will help ABC Institution decide on the subject matter.

We thank you for the opportunity given to us to be of service in your development scheme.

Thank you.

Yours faithfully,
For: Brown and Co Associates

A. Jackson (mniqs)
(Principal Partner)

2.4 QUANTITY SURVEYOR'S ROLE IN PUBLIC AND PRIVATE SECTORS

A Quantity Surveyor (QS) is one of the relevant members of the design team. A QS is a cost economist, advising clients and Architects on the probable cost of alternative designs when necessary. His recommendation enhances design and construction regarding pre-determined limits of expenditure. The QS advises on methods for arranging building and engineering contracts. A QS also prepares the contract bill of quantities and, where suitable, negotiates construction contracts with contractors (Seeley, 1997). Others are preparing an estimate of final costs and valuations for payments, in charge of measurement and valuation of variations, preparing the contractor's final account, etc. The Quantity Surveyor's role cannot be overemphasised in the public and private sectors of the construction industry.

2.5 QUANTITY SURVEYOR IN CONTRACTING ORGANISATIONS

The services of thc Quantity Surveyors engaged in contracting organisations tend to be large in scope with the smaller firms but rather more specialised with the larger organisations (Seeley, 1997). The contractor Quantity Surveyor's activities include contract bill of quantities preparation for small construction projects, market survey from which the contractor can prepare future cost projections, and materials schedule preparation. Others are compiling target figures, planning contracts, preparing progress charts, and applying to the Architect for variation orders if drawings or site instructions fluctuate the work. Others are concurring the value of variations and Sub-contractor accounts and comparing the costs of alternative methods of site operations so that the most economical can be adopted.

2.6 APPOINTMENT OF QUANTITY SURVEYORS AND PROFESSIONAL FEES

The appointment, duties, and responsibility of the Quantity Surveyors in connection with construction contracts are treated under the following sub-headings:

i. **Appointment**: Experience shows that a Quantity Surveyor is appointed either through the Architect or directly by the client. The Nigerian Institute of Quantity Surveyors emphasises the importance of employers' direct quantity surveying appointments. This is to guarantee independent cost advice on the project. There are instances where the Architects may recommend a Quantity Surveyor to the client. The long-standing working experience between the Architect and the Quantity Surveyor is among the factors determining this.
ii. **Scope of Work**: This section describes the engaged Quantity Surveyor's scope of work. The Quantity Surveyor's scope of duty should be explicit (pre-contract, post-contract, or both). The consultancy services agreement will be signed by the Quantity Surveyors with the client, or the Architect can

sign it on his behalf with the client. A commission letter must clearly state the type of duties to be performed by the Quantity Surveyor before the commencement of work

iii. **Remuneration**: The consultants shall be paid for the services rendered. This is in accordance with the scale of fees approved by the Federal Government of Nigeria and published under the title "Consultancy fee payable by the public sector." The fee is based on the accepted estimated cost of the project but recalculated on the accepted tendered cost of construction in the lesser of the two amounts. This may also include the final cost of the provisional sum within the estimated project cost. Fees are based on the accepted total cost of the contract, regardless of the number of construction contracts involved, in accordance with the employers' requirements. Where repetitive work is involved, fee calculations for each element are made; full fees are allowed on non-repetitive work such as preliminaries and each type's first building or element. Fees indicated on the remainder of the sliding scale for repetitive work shall then be applied. Where full services are not required regarding the design and build construction contract or where standard drawings are utilised, the fees shall further be reduced on a percentage basis of the actual services to be performed as negotiated between the employer and the consultant. Reimbursement expenses shall be allowed in addition to the agreed fee when authorised and approved by the employer regarding transportation, accommodation, and substance in conjunction with travel, resident supervisor, and the printing of documents. Fees shall be paid in Nigerian naira upon completing each stage of work. In special circumstances, the employer may approve interim payments within the period of work done. The total fees are payable to the consultant in naira (N) based on the estimated total cost (ETC). If the tendered value is less than the ETC, the fees shall be recalculated based on the accepted tender.

iv. **Method of Payment**: It is usually agreed that sums due to the Quantity Surveyor shall be paid in naira. The fee shall be due upon completion of each stage as set out in the agreement.

v. **Additional Services**: Additional services carried out as requested by the Quantity Surveyor shall be undertaken on such terms and conditions as may be agreed upon among the parties. Payment to the consultant for extra services shall be the lump sum agreed upon between the Quantity Surveyor and the employer.

vi. **Facilities**: The Consultant Quantity Surveyor shall provide enough specialised staff to undertake the work. The work should be done with the approval of the client.

vii. **Commencement of Services**: The contract shall become effective upon the consultant's receipt of employer's written notice to proceed with the services.

viii. **Time of Completion**: The Consultant Quantity Surveyor services shall be completed on the completion of the pre-contract duties and post-contract duties, or both.

ix. **Care, Diligence, and Responsibilities**: The Consultant Quantity Surveyor shall diligently perform their duties. The QS shall carry out the agreement with all reasonable expedition and dispatch.

x. **Non-Assignment**: The Consultant Quantity Surveyor shall not have the right to transfer the benefits and objectives of his agreement or any part thereof.

xi. **Postponement or Abandonment**: The Quantity Surveyor shall be paid for services rendered regarding postponed, abandoned, or delayed parts of the works.

xii. **Report on Progress of Work**: A monthly report on work progress shall be submitted to the employer. The monthly report comprises minutes of the site meeting, a monthly certificate of payment, records of labour on-site, and progress photographs where applicable.

xiii. **Arbitration:** The Arbitrator shall settle matters that arise between the parties which cannot be settled mutually. The parties shall refer the matter to the agreed Arbitrator. The Arbitrator's findings shall be absolute and mandatory on the parties hereof.

xiv. **Termination of the Agreement**: The agreement shall terminate on the date the final payment is made to the Quantity Surveyor.

Professional Scale of Fees for Construction Consultants (Special Features of the Federal of Ministry Works and Housing Revised Edition, April 1996)

Note: There is on-going consultation to harmonise the scale of fees prepared by respective professional institutions.

1. **Prime Consultant Fee**
   The prime consultant is the head of the consultants' project team. They initiate the design and generate the sketches and coordinate and manage the construction project in some instances.
2. **Repetitive Works**
   This is reduced, as shown on the scale. The reduction shall not apply to single buildings but to substructure work and post-contract services.
3. **Man-Month Rate**
   To be received annually with the CBN inflation index for that year. Professionals in the built environment widely use this method. This is because the method is to their advantage.
4. **Project Manager**
   The Quantity Surveyor could be a Project Manager, and the fee for such services is 1%.
5. **Reimbursable Expenses**
   Maximum of 1% of the project cost and is distributed as follows:
   - Architect or Prime Consultant 40%
   - Structural Engineer 15%
   - Quantity Surveyor 15%
   - Mechanical Engineer 10%
   - Electrical Engineer 10%

- Geotechnical 5%
- Builders 5%

Note: Where 1% of the project's cost is inadequate, approval should be sought from the client before a case for additional reimbursement is made. Where Geotechnical/Builders are absent, their share goes to the prime consultant.

6. **Additional Services**

   Stage 1: Feasibility Studies: 0.20% of the ETC.
   Stage 2: Pricing Bill of Quantities for a suspended, abandoned, or terminated contract—0.1% of ETC.
   Stage 3: Making any site visit/site meeting other than the normal single monthly meeting will be compensated using the man-month rate of the time change.

Note: Normal supervision beyond the original contract period will also be compensated.

7. **Other Additional Charges**
   i. Schedule of material—0.1% of ETC
   ii. Litigation or Arbitration—Negotiable
   iii. Rehabilitation of damaged existing buildings—Based on the scale of fees
   iv. Construction Cost of Replacement for Insurance and other purposes—0.5% of the cost
   v. Independent QS advice on the issue of liquidation will be remunerated at the man-month rate.
   vi. Fluctuation (prime investigation)—maximum of 1% of the cost
   vii. Resident Supervision by QS—Remunerated by man-month rate.

8. **Stage Payment**

   Stage1—25% of fees of the ETC of the project (up to the preliminary estimated stage)
   Stage 2—50% of fees on ETC (tender report/signing of the contract).
   Stage 3—25% of fees based on the (post-contract) total construction cost.

   Fees for stages 1 and 2 shall be determined based on the project's ETC. Fees for stage 3 shall be determined based on the ETC of the project, which is reviewed regularly. The total construction sum shall only be utilised to modify the final payment in stage 3. Where the ETC is known, the tender sum shall be employed.

9. **Multi-various Projects**

   There is a need to make multifarious projects clearer and more comprehensible. The essence is to ensure that the client gets value for the money spent under this scheme. The following are the attributes that qualify a project to fall into this category:

- Variety of uses
- Substantial infrastructure

Examples:

- Substation and power distribution plant
- Sewers and sewage treatment basement plant
- Central water supply distribution system
- Road network, etc.

Therefore, a good example of multifarious projects includes the following:

- Teaching/specialist central hospital complex
- Major residential housing estate (at least five hectares of land)
- Industrial layout complex (project cost for fee computation shall exempt the equipment cost).

The following are the steps to calculate fees for the multifarious project:

- Determine the total cost of the project.
- Determine the cost of non-repeat and repeat projects, if any.
- Determine the cost of the following infrastructure components:
  i. Residential and office buildings
  ii. Civil/structure infrastructures
  iii. Electrical installations
  iv. Mechanical installations, etc.

The areas to which this can be applied are:

a. Prime consultancy applies only to the area of specialisation and fees.
b. Non-prime consultancy fees shall be in accordance with the normal scale of fees.
c. Structural/civil parts necessary for mechanical and electrical works shall be treated along with civil/structural works.

Note: Students are advised to get an approved revised edition (1996) of the professional scale of fees for construction consultants in Nigeria.

Example I shows a sample of fee computation for a multifarious project.

Example I

Project Profile

i. Project residential
ii. Area: 6.5 hectares
iii. Total project cost: ₦500,000,000
iv. Number of buildings: 9

Infrastructure Works

- Civil: ₦40,000,000
- Mechanical: ₦25,000,000
- Electrical: ₦35,000,000

=₦**100,000,000**

Building Works (Housing)

i. 1 no. prototype
substructure of non-repeated works: 230,000,000
ii. repeated works (superstructure of 8 nos) 170,000,000

Fee computation for Infrastructure Works

i. Civil Engineer's fees as a Prime Consultant.

Cost of works		= ₦40,000,000=
First	₦5,000,000 @ 4.75%	= 237,500 =
Next	₦10,000,000 @ 4.50%	= 450,000 =
Next	₦15,000,000 @ 4.25%	= 637,000 =
Balance	₦10,000,000 @ 4.00%	= 400,000 =
	₦40,000,000	**₦1,725,500** =

ii. Quantity surveyor's fees when appointed where the civil engineer is prime consultant.

Cost of works		= ₦40,000,000 =
First	₦5,000,000 @ 1.37%	= 68,500 =
Next	₦10,000,000 @ 1.225%	= 125,000 =
Next	₦15,000,000 @ 1.5%	= 172,000 =
Balance	₦10,000,000 @ 1.00%	= 100,000 =
	₦40,000,000	**₦466,000 =**

iii. Mechanical engineer's fees as prime consultant

Cost of works		= ₦25,000,000=
First	₦5,000,000 @ 4.75%	= 237,500 =
Next	₦10,000,000 @ 4.50%	= 450,000 =
Balance	₦10,000,000 @ 4.25%	= 425,000 =
	₦25,000,000	**₦1,112,500**

iv. Quantity surveyor's fees when appointed where the mechanical engineer is prime consultant

Cost of work		= ₦25,000,000 =
First	₦5,000,000 @ 2.75%	= 137,500 =
Next	₦10,000,000 @ 2.50%	= 250,000 =
Balance	₦10,000,000 @ 2.30%	=230,000 =
	₦25,000,000	**₦617,500=**

v. Electrical engineer's fees as prime consultant

Cost of work		= ₦35,000,000 =
First	₦5,000,000 @ 4.75%	= 237,500=
Next	₦10,000,000 @ 4.50%	= 450,000=
Next	₦15,000,000 @ 4.25%	= 63,7500=
Balance	₦5,000,000 @ 4.000%	= 200,000=
	₦35,000,000	**₦1,525,000=**

vi. Quantity surveyor's fees when appointed where the electrical engineer is prime consultant.

Cost of work		= ₦35,000,000=
First	₦5,000,000 @ 2.75%	= 137,500 =
Next	₦10,000,000 @ 2.50%	= 250,000=
Next	₦15,000,000 @ 2.30%	= 345,000 =
Balance	₦5,000,000 @ 2.000%	= 100,000 =
	₦35,000,000	**832,5000 =**

Fees Computation for Building Works

i. Architect's fees as prime consultant

a. Cost of non-repeated works = 230,000,000=

First	₦5,000,000 @ 4.75%	= 237,500
Next	₦10,000,000 @ 4.50%	= 450,000=
Next	₦15,000,000 @ 4.25%	= 637,000=
Next	₦45,000,000 @ 4.00%	= 1,800,000=
Next	₦75,000,000 @ 3.50%	= 2,625,000=
Balance	₦80,000,000 @ 3.00#	= 2,400,000
	₦230,000,000	**₦8,150,000 =**

b. Cost of repeated works = N170,000,000

First	₦5,000,000 @ 4.75%	= 237,500
Next	₦10,000,000 @ 4.50%	= 450,000
Next	₦15,000,000 @ 4.25%	= 637,000
Next	₦45,000,000 @ 4.00%	= 1,800,000
Next	₦75,000,000 @ 3.50%	= 2,625,000
Balance	₦20,000,000 @ 3.00%	= 600,000
	₦170,000,000	**₦6,350,000**

Repeat factor (₦6,350,000.00) @ 30% = ₦1,905,000=

Summary

a = ₦8,150,000
b = ₦1,905,000
₦10,055,000

Stages I and II = ₦10,055,000 @ 75% = ₦7,541,250=
Stage III = ₦10,055,000 @ 25% = ₦2,513,750 =

ii. Structural engineer's fees

a. Cost of non-repeated works = **₦230,000,000=**
First ₦5,000,000 @ 3.00% = 150,000 =
Next ₦10,000,000 @ 2.50% = 250,000=
Next ₦15,000,000 @ 2.25% = 337,000 =
Next ₦45,000,000 @ 2.00% = 900,000 =
Next 75,000,000 @ 1.75% = 1,312,000 =
Balance ₦80,000,000 @ 1.50% = 1,200,000=
₦230,000,000 **₦4,150,000=**

b. Cost of repeated works = **₦170,000,000**
First ₦5,000,000 @ 3.000% = 150,000
Next ₦10,000,000 @ 2,00% = 250,000
Next ₦15,000,000 @ 2.25% = 337,000
Next ₦45,000,000 @ 2.00% = 900.000
Next ₦75,000,000 @ 1.75% = 1,312,000
Balance ₦20,000,000 @ 1.50% = 300,000
₦170,000,000 **₦3,250,000**
Repeat factor (₦3,250,000) @ 30% = ₦975,000 =
Summary
a = ₦4,150,000
b = ₦975,000
₦5,125,000
Stages I and II = ₦5,125,000 @ 75% = **₦3,843,750** =
Stage III = ₦5,125,000 @ 25% = **₦1,281,250**

iii. Mechanical engineer's fees

a. Cost of non-repeated works = ₦230,000,000 =
First ₦5,000,000 @ 1.95% = 97,500 =
Next ₦10,000,000 @ 1.75% = 175,000 =
Next ₦15,000,000 @ 1.55% = 232,500 =
Next ₦45,000,000 @ 1.35% = 607,500 =
Next ₦75,000,000 @ 1.15% = 862,000 =
Balance ₦80,000,000 @ 1.00% = 800,000 =
₦230,000,000 **₦2,774,500 =**

b. Cost of repeated works = ₦170,000,000 =
First ₦5,000,000 @ 1.95% = 97,500 =
Next ₦10,000,000 @ 1.75% = 175,000 =
Next ₦15,000,000 @ 1.55% = 232,500 =
Next ₦45,000,000 @ 1.35% = 607,500 =
Next ₦75,000,000 @ 1.15% = 862,000 =
Balance ₦20,000,000 @ 1.00% = 200,000 =
₦170,000,000 **₦2,174,500 =**

Repeat factor (₦2,174,500) @ 30% = ₦652,350 =
Summary
a = ₦2,774,500 =
b = ₦652,350=
₦3,426,850 =
Stages I and II = ₦ 3,426,850 @ 75% = **₦2,570,137:50k**
Stage III = ₦3,426,850 @ 25% = **₦856,712:50k**

iv. Electrical engineer's fees

a. Cost of non-repeated works = **₦230,000,000 =**

First	₦5,000,000 @ 1.95%	= 97,500 =
Next	₦10,000,000 @ 1.75%	= 175,000 =
Next	₦15,000,000 @ 1.55%	= 232,500 =
Next	₦45,000,000 @ 1.35%	= 607,500 =
Next	₦75,000,000 @ 1.15%	= 862,000 =
Balance	₦80,000,000 @ 100%	= 800,000 =
	₦230,000,000	**₦2,774,500 =**

b. Cost of repeated works = ₦170,000,000 =

First	₦5,000,000 @ 1.95%	= 97,500 =
Next	₦10,000,000 @ 1.75%	= 175,000 =
Next	₦15,000,000 @ 1.55%	= 232,500 =
Next	₦45,000,000 @ 1.35%	= 607,500 =
Next	₦75,000,000 @ 1.15%	= 862,000 =
Balance	₦20,000,000 @ 1.00%	= 200,000 =
	₦170,000,000	**₦2,174,500 =**

Repeat factor (₦2,174,500) @ 30% = ₦652,350 =
Summary
a = ₦2,774,500 =
b = ₦652,350=
₦3,426,850 =
Stages I and II = ₦ 3,426,850 @ 75% = **₦2,570,137:50k**
Stage III = ₦3,426,850 @ 25% = **₦856,712:50k**

v. Quantity surveyor's fees

a. Cost of non-repeated works = ₦230,000,000 =

First	₦5,000,000 @ 2.75%	= 137,500
Next	₦10,000,000 @ 2.50%	= 250,000 =
Next	₦15,000,000 @ 2.30%	= 345,000 =
Next	₦45,000,000 @ 2.00%	=900,000 =
Next	₦75,000,000 @ 1.75%	= 1,312,500 =
Balance	₦80,000,000 @ 1.40%	= 1,120,000 =
	₦230,000,000	**₦4,064,500 =**

b. Cost of repeated works = ₦170,000,000 =

First	₦5,000,000 @ 2.75%	= 137,500
Next	₦10,000,000 @ 2.50%	= 250,000 =
Next	₦15,000,000 @ 2.30%	= 345,000 =
Next	₦45,000,000 @ 2.00%	= 900,000 =
Next	₦75,000,000 @ 1.75%	= 1,312,500 =
Balance	₦20,000,000 @ 1.40%	= 280,000 =
	₦170,000,000	**₦3,225,000** =

Repeat factor (₦3,225,000) @ 30% = ₦967,500 =

Summary

a = ₦4,064,500

b = ₦967,500

₦5,032,00

Stages I and II = ₦ 5,032,000 @ 75% = **₦3,774,000**

Stage III = ₦5,032,000 @ 25% = **₦1,258,000** =

vi. Project manager's fees

Cost of project = ₦500,000,000 =

First	₦5,000,000 @ 1.00%	= 50,000
Next	₦10,000,000 @ 0.90%	= 90,000 =
Next	₦15,000,000 @ 0.80%	= 120,000 =
Next	₦45,000,000 @ 0.70%	=315,000 =
Next	₦75,000,000 @ 0.60%	= 450,000 =
Next	₦150,000,000 @ 0.50%	= 750,000=
Balance	₦200,000,000 @ 0.40%	= 280,000 =
	₦500,000,000	**₦2,575,000** =

Stages I and II = ₦2,575,000 @ 75% = **₦1,931,250** =

Stage III = ₦2,575,000 @ 25% = **₦634,750** =

APPENDIX 1

Relevant Tables Extracted From the Professional Scale of Fees for Construction Consultants

TABLE A.1 Prime Consultant Scale of Fees

Cost of Project	Fees Payable as a Percentage of the Cost of the Project
Up to 5 million	4.75
Next 10 million or part thereof	4.50
Next 15 million or part thereof	4.25
Next 45 million or part thereof	4.00

(Continued)

TABLE A.1 (Continued)

Cost of Project	Fees Payable as a Percentage of the Cost of the Project
Next 75 million or part thereof	3.50
Next 150 million or part thereof	3.00
Next 200 million or part thereof	2.50
Balance over 500 million	1.75

TABLE A.2 Consultants Scale of Fees for Repetitive Works

Cost of Project	Fees Payable as a Percentage of the Cost of the Project
Initial project	100
Next 10 repetitions or part thereof	30
Next 10 repetitions or part thereof	20
Next 10 repetitions or part thereof	15
Next 10 repetitions or part thereof	12.5
Next 10 repetitions or part thereof	10
Next 10 repetitions or part thereof	7.5
Balance over 500 repetitions	5

Source: Please, single building, substructure, and post-contract services are exempted from the repetitive deduction.

TABLE A.3 Scale of Fees Where Project Manager or any Other Consultant Is the Prime Consultant

Cost of Project	Fees Payable as a Percentage of the Cost of the Project
Up to 5 million	1.00
Next 10 million or part thereof	0.90
Next 15 million or part thereof	0.80
Next 45 million or part thereof	0.70
Next 75 million or part thereof	0.60
Next 150 million or part thereof	0.50
Next 200 million or part thereof	0.40
Balance over 500 million	0.30

TABLE A.4 Scale of Fees Where the Prime Consultant Is Not the Architect

Cost of Project	Fees Payable as a Percentage of the Cost of the Project
Up to 5 million	3.00
Next 10 million or part thereof	2.50
Next 15 million or part thereof	2.25
Next 45 million or part thereof	2.0
Next 75 million or part thereof	1.75
Next 150 million or part thereof	1.5
Next 200 million or part thereof	1.25
Balance over 500 million	1.00

Notes: Table 4 is an application to:

i. Scale of fees for structural engineers where the prime consultant is the architect.
ii. Scale of fees for structural engineers where the prime consultant is the electrical or mechanical engineer.

TABLE A.5 Scale of Fees for Mechanical and Electrical Engineering Works Where the Prime Consultant Is the Architect or Structural Engineer

Cost of Project	Fees Payable as a Percentage of the Cost of the Project
Up to 5 million	1.95
Next 10 million or part thereof	1.75
Next 15 million or part thereof	1.55
Next 45 million or part thereof	1.35
Next 75 million or part thereof	1.15
Next 150 million or part thereof	1.0
Next 200 million or part thereof	0.85
Balance over 500 million	0.65

TABLE A.6 Scale of Fees for Quantity Surveying Works in Projects Where the Prime Consultant is the Architect, Mechanical, Electrical, or Structural Engineer

Cost of Project	Fees Payable as a Percentage of the Cost of the Project
Up to 5 million	2.75
Next 10 million or part thereof	2.50
Next 15 million or part thereof	2.30

(*Continued*)

Table A.6 (Continued)

Next 45 million or part thereof	2.0
Next 75 million or part thereof	1.75
Next 150 million or part thereof	1.40
Next 200 million or part thereof	1.00
Balance over 500 million	0.80

TABLE A.7 Scale of Fees for Quantity Surveying Works in Projects Where the Prime Consultant Is the Civil Engineer

Cost of Project	Fees Payable as a Percentage of the Cost of the Project
Up to 5 million	1.37
Next 10 million or part thereof	1.25
Next 15 million or part thereof	1.15
Next 45 million or part thereof	1.00
Next 75 million or part thereof	0.87
Next 150 million or part thereof	0.70
Next 200 million or part thereof	0.50
Balance over 500 million	0.40

Note: he scale of fees for quantity surveying works in projects like roads and bridges, where the Prime Consultant is the Civil Engineer, is presented in Table A.7.

Also, in a multifarious project where there are several prime consultants, for example, university complexes or housing estates, among others, the Civil Engineer (Prime Consultant) takes charge of estate roads, the Architect (Prime Consultant) takes charge of the building, the Electrical Engineer (Prime Consultant) takes charge of street lighting, and the Structural Engineer (Prime Consultant) takes charge of the sewage treatment plant. Under these circumstances, the scale of fees for Quantity Surveyors as in Tables A.6 and A.7 is applied appropriately, using each Prime Consultant as a base to determine the ETC for his area of operation.

However, where the Quantity Surveyor employs the services of the Mechanical or Electrical Engineer in the bill preparation for a project or making an input into the preparation of the BOQ for a construction project, the fee payable to each Engineer shall be negotiated. The negotiation should be less than 50% of the fee accruable to the QS. However, where the service of the Civil Engineer is engaged by the QS in the preparation of the BOQ for a construction project or in making an input into the bill preparation for a project, the fee payable to the Civil Engineer shall be negotiated. Similarly, the negotiation should be less than 50% of the fee accruable to the QS.

Note: The rules enunciated in the last paragraph shall apply to multifarious projects appropriately.

Assignment

You are the branch manager of a firm that has been appointed as Quantity Surveyor for a university project. Draft a letter for your head office telling them the fee to be expected. The brief is as follows:

i. Hall of residence—1 No—N15,500,000=
ii. 10 Nos identical staff houses: N5,000,000= each
iii. 20 Nos identical blocks of classrooms–N8,000,000= each
iv. The substructure is 12.5% of the total cost of building.
v. University roads + drainage: N40,000,000
vi. Assume maximum reimbursement for reproduction, travelling, etc.

2.7 COMPARISON BETWEEN DEVELOPED AND DEVELOPING COUNTRIES PROFESSIONAL QUANTITY SURVEYING PRACTICES

There is no doubt that a Quantity Surveyor is a key professional in the developed and developing countries' construction industries. It is a profession that deals with construction costs and contributes to the project's success regarding quality, cost, and time. The QS as a professional, irrespective of the country, ensures that construction resources are utilised to the benefit of society via construction cost consultancy and financial management of projects to clients and other design team members during the pre- and post-contract process of the project (Owusu-Manu *et al.*, 2014; Yogeshwaran *et al.*, 2018). The chapter emphasises that, besides the inadequacies in certain competencies in developing countries quantity surveying practices, thereby failing to reach the expected standard in some construction projects, the QS's role irrespective of the country ought to be the same. This may have enhanced unethical practices in developing countries' construction industries. Thus, there are insignificant differences between the roles of a QS in developed and developing countries. The digitalisation trend has been infused into the quantity surveying profession, expanding the role beyond the traditional roles and duties in developed and developing countries, especially in developed countries such as the United Kingdom. Professional bodies such as the Royal Institution of Chartered Surveyors (RICS) (2015) emphasised the need for a QS graduate to attain and demonstrate these proficiencies. The Institute based the competencies on technical and mandatory competencies. In some developing countries, the professional bodies are lax concerning competencies and, by extension, affect the performance of Quantity Surveyors in practice. The gap between the industry and the quantity surveying graduates regarding the expected minimum expected competencies has not helped matters. It has increased the industry's dissatisfaction about Quantity Surveyors competencies of present graduates (Perera *et al.*, 2011), especially in developing countries. Many higher education institutions in developing countries believe that their graduates must focus on competencies as documented by their regulatory bodies without considering and consulting with the industry's needs. The outcome has increased the unemployability of Quantity Surveying graduates in many developing countries. Nigeria is not exempted.

2.8 COMMUNICATION STYLES RELEVANT TO THE QUANTITY SURVEYING PROFESSION

Communication style within and outside the project team is key in the construction industry. A communication style can be described as how one verbally and non-verbally interrelates to signal how the correct meaning should be construed or understood. Norton (1983) categorised communication styles into ten different types. This includes precise, friendly, open, attentive, relaxed, impression-leaving, animated, contentious, dramatic, and dominant. McCallister (1992) clustered communication styles into Socratic, reflective, and noble. The noble style is straightforward and directive (Dasgupta *et al.*, 2013). Thus, the way a QS communicates with others, verbally and non-verbally, is a function of how other stakeholders receive and interpret verbal messages. The role of QS in the construction industry makes it pertinent to be sufficiently competent in communication skills. It would enhance the progress of the built environment because the role is expanding and might include quality assurance coordination, supply chain management, sustainable construction management, life cycle costing, facilities management, and development appraisal in the future (Karunarathna, 2006; Yogeshwaran *et al.*, 2018). Hence, the QS professional bodies emphasised communication skills as a component of the expected competencies to ensure that practicing Quantity Surveyor demonstrate the highest standards of professional brilliance (Badu and Amoah, 2004; RICS, 2015). The communication skills are germane for the QS. It would promote the QS to gain competence and survive in a competitive industry. The essence of training in higher education institutions is to enhance employability. Enhancing employability requires generic skills, such as communication skills. Communication skills was identified as a component of the generic skills that construction graduates, including quantity surveying graduates, should acquire to enhance their employability in the labour market (Aliu and Aigbavboa, 2021; Ebekozien *et al.*, 2021, 2022).

2.9 SUMMARY

This chapter covered the formation of a quantity surveying firm and the types of practice. This is followed by the design process and the professionals involved, including their cultural differences and work habits. The chapter also covers the reporting of tenders, the role of a Quantity Surveyor in pre- and post-contract administration, and the comparison between developed and developing countries' professional QS practices. The chapter highlights different communication styles relevant to enhancing a QS profession. The chapter concluded with the appointment of a Quantity Surveyor and professional fees for consultants in the construction industry. The next chapter gives an in-depth description of professional ethics, the roles of regulatory bodies, the institute membership examination, and the conduct of a quantity surveying practice.

REFERENCES

Adunfe, A.A., Moses, E.O., Olalekan, A.H. and Olalekan, S.A. 2022. Assessment of the innovation performance of Nigerian quantity surveying firms. *International Journal of Innovative Research and Development, 11*(3), pp. 11–19.

Aliu, J. and Aigbavboa, C.O. 2021. Key generic skills for employability of built environment graduates. *International Journal of Construction Management*. https://doi.org/10.1080/15623599.2021.1894633.

Ashworth, A., Hogg, K. and Higgs, C. 2013. *Willis's Practice and Procedure for the Quantity Surveyor*. West Sussex: John Wiley & Sons.

Badu, E. and Amoah, P. 2004. Quantity surveying education in Ghana. *The Ghana Engineer*, pp. 1–12. Available at: www.icoste.org/GhanaEdu.pdf (accessed 30 May 2022).

Chappell, D. and Dunn, M.H. 2015. *The Architect in Practice*. London: John Wiley & Sons.

Dasgupta, S.A., Suar, D. and Singh, S. 2013. Impact of managerial communication styles on employees' attitudes and behaviours. *Employee Relations, 35*(2), pp. 173–199. https://doi.org/10.1108/01425451311287862

Ebekozien, A., Aigbavboa, C., Aliu, J. and Thwala, W.D. 2022. Generic skills of future built environment practitioners in South Africa: Unexplored mechanism via students' perception. *Journal of Engineering, Design and Technology*. https://doi.org/10.1108/JEDT-10-2021-0571.

Ebekozien, A., Aigbavboa, C., Thwala, W.D., Aigbedion, M. and Ogbaini, I.F. 2021. An appraisal of generic skills for Nigerian built environment professionals in workplace: The unexplored approach. *Journal of Engineering, Design and Technology*. https://doi.org/10.1108/JEDT-09-2021-0453.

Hughes, W.P. 2003. A comparison of two editions of the RIBA plan of work. *Engineering, Construction and Architectural Management, 10*(5), pp. 302–311. https://doi.org/10.1108/09699980310502919

Karunarathna, A.M.A.P. 2006. Five years look ahead: Quantity surveyors' expectations in Sri Lanka. Unpublished bachelor's dissertation, University of Moratuwa.

Male, S. 1990. Professional authority, power and emerging forms of 'profession' in quantity surveying. *Construction Management and Economics, 8*(2), pp. 191–204. https://doi.org/10.1080/01446199000000016

McCallister, L. 1992. *I Wish I'd Said That: How to Talk Your Way Out of Trouble and into Success*. New York: John Wiley and Sons.

Musa, N.A., Oyebisi, T.O. and Babalola, M.O. 2010. A study of the impact of information and communications technology (ICT) on the quality of quantity surveying services in Nigeria. *The Electronic Journal of Information Systems in Developing Countries, 42*(1), pp. 1–9.

Norton, R. 1983. *Communicator Style: Theory, Applications and Measures*. Beverly Hills, CA: Sage.

Owusu-Manu, D.-G., Edwards, D.J., Holt, G.D. and Prince, C. 2014. Industry and higher education integration: A focus on quantity surveying practice. *Industry and Higher Education, 28*(1), pp. 27–37. https://doi.org/10.5367/ihe.2014.0188

Perera, S., Pearson, J., Robson, S. and Ekundayo, D. 2011. Alignment of academic and industrial development needs for quantity surveyors: The views of industry and academia. In *Proceedings of the RICS COBRA Research Conference*, pp. 676–686. Available at: www.academia.edu/901056/Alignment_of_academic_and_industrial_development_n eeds_forquantity_surveyors_the_views_of_industry_and_academia (accessed 10 June 2022).

Peter, O.O., Eze, E.C. and Anthony, A.A. 2019. Assessment of quantity surveying firms' process and product innovation drive in Nigeria. *SEISENSE Journal of Management*, *2*(2), pp. 22–38.
Royal Institute of British Architects (RIBA). 1887. *Journal of Proceedings of the Royal Institute of British Architects*, *4*(1).
Royal Institute of British Architects (RIBA). 1963. *Plan of Work for Design Team Operation*. London: RIBA.
Royal Institute of British Architects (RIBA). 1972. *Architect's Appointment*. London: RIBA.
Royal Institute of British Architects (RIBA). 1997. *Plan of Work for Design Team Operation*. London: RIBA Publications.
Royal Institution of Chartered Surveyors (RICS) (2015). Member (MRICS). Available at: www.rics.org/lk/join/member-mrics/ (accessed 6 June 2022).
Seeley, I.H. 1997. *Quantity Surveying Practice*. London: Macmillan.
Yogeshwaran, G., Perera, B.A.K.S. and Ariyachandra, M.R.M.F. 2018. Competencies expected of graduate quantity surveyors working in developing countries. *Journal of Financial Management of Property and Construction*, *23*(2), pp. 202–220. https://doi.org/10.1108/JFMPC-06-2017-0019

3 Professional Ethics

3.1 ETHICS BACKGROUND

Every profession lays down ethics that members of the professional body should obey. A professional is a person who has acquired skills and is trained to apply a specific body of established knowledge and practices to practical issues. A professional is expected to register with the appropriate body. Every professional body has a code of ethics guiding members of the body. The main components of professionalism include independence of judgement, dedication to the public interest, and a high standard of expertise. Professionalism is a set of behaviours such as competence, delivery of valuable services, consultancy, honesty, and fairness. This is the act where the professional conducts his duties and obligations. A professional can only attain this feat by understanding where to draw a line between wrong and right as it applies to his profession (RICS, 2020). He must understand the ethics guiding his profession. Ethics is defined as the science of morals, moral principles, or codes. Several authors have attempted to define ethics as it applies to their field. Quantity surveying is not an exemption. To some authors, ethics is the science of moral values and duties. Ethics are the checkmate of human conduct in directing what should be done and what one should be kept away from. While others believe that ethics is all about the theoretical study of what is right or wrong in humans, ethical behaviour originally and essentially stems from a person's value system. This is likely to know right and wrong and have the bravery to do what is right. Ethics is all about no right way to do what is wrong. However, society has become desensitised to unethical behaviour in every sector, including politics. There is an allegation that some politicians buy and sell construction contracts (Ebekozien, 2020). This allegation should be investigated.

Professional ethics is the practice of moral principles or rules of conduct that define occupational ethics. This explains the suitability of an abstract standard of behaviour for practical undertakings. This is not solely restricted to advanced technologies, exchanges, events, pursuits, and evaluations of establishments. This incorporates more pragmatic conceptualisation and public expectations in the interest of responsibility, willingness to serve the public, and astute capabilities. This agrees with the definition of professional ethics by Fan *et al.* (2001). They opined that giving one's ideal guarantees that an employer's intrigue is appropriately thought about. Thus, professional values are demonstrated in ethical codes. The ethical codes refine quantity surveying profession practices. Ethics is a flexible tent. Ethical standards are active issues. Certain actions may be ethical today or in a society and certain situations, but they may be viewed otherwise by others or in another situation. Thus, appraising behaviour to keep pace with changing standards continuously is germane. An expert's ethical scene comprises decisions, choices, and the commonsense utilisation of shared fundamental beliefs and standards. A key issue in this

DOI: 10.1201/9781032661391-3

methodology is the need for practitioners to consider their proficient practice. Also, consider their decisions against the codes of practice created within the profession (RICS, 2020).

The crucial standards for most professional bodies are integrity, objectivity, competence and care, secrecy, and conduct. Even though the need for various qualities and rules may differ between professions, the overall list tends to be held in common. An additional significant professional right is named the "right of conscientious refusal." When compelled by the boss or employer, a worker can decline to participate in dishonest conduct. This may happen in work or non-work environments. Careful rebuttal may be done by simply not participating in the activity one sees as morally wrong. Behaving ethically is at the heart of what it means to be a professional. It is a core value of a professional (Yap *et al.*, 2022). This is key because the industry is vulnerable to corrupt practices such as collusion, kickbacks, nepotism, bribery, and corruption (Yap *et al.*, 2020; Ebekozien, 2020). Bowen *et al.* (2012) believe the possibility of engaging in unethical practices can be explained using the "fraud triangle" theory. This comprises the three basics of opportunities (forms of corruption), pressure for corruption, and rationale for corruption. To bridge this gap, professional norms should be learned through active involvement and the experience of progressing in the job. To improve ethical compliance in the construction industry, professional practice curricula for professional education and continuing professional development (CPD) courses should be introduced during ethical training to improve ethical dilemmas during practice (Bowen *et al.*, 2015).

3.2 PROFESSIONAL CONDUCTS

Like other professionals in the built environment, Quantity Surveyors acknowledge that they shoulder special responsibilities. Quantity surveying is a profession rooted in professional ethics and ethical values. These responsibilities should conform to ethical standards from a layperson's perspective. The core values of quantity surveying include integrity, transparency, accountability, honesty, human dignity, autonomy, and justice. The core ethical values are shared within the built-environment community. Quantity surveying is a professional practice with regulations, rules of conduct, and ethical codes upheld by established professional bodies to enhance assurance and project clients against substandard practices (RICS, 2018, 2020). The Constitution, bye-laws, code of professional conduct, and general laws of Nigeria govern members of the Nigerian Institute of Quantity Surveyors (NIQS). Decree No. 31 of 1986 set up the Quantity Surveyors Registration Board of Nigeria (QSRBN) and regulated the practice of the quantity surveying profession in Nigeria (NIQS, 1998). The code aims to regularise the level of discipline, skill, and behaviour of members of the Institute and the registration board in fulfilling the practice of the profession in Nigeria. Violating any of the Clauses in the bye-laws controlling and regulating the Quantity Surveyor in the code will subsequently affect the practitioner's membership in the Institute (NIQS). In the case of practicing firms, the principal/partner of the firm shall be vicariously liable for any contravention of the code committed by any staff member in their official duties, always provided that it is established beyond a reasonable doubt that the staff does not commit the tortuous act. This misrepresents the good intention of the partnership or acts in his selfish interest.

Therefore, it is necessary to discuss the main duty of the Quantity Surveyor to appreciate professional conduct and misconduct. The main duty of the Quantity Surveyor in the construction industry is to guarantee that the assets of the industry are used to the best favourable position of the society by providing intermediate financial management for construction projects and cost consultancy services to the employer and Architect during the development process (Seeley, 1997; Ashworth *et al.*, 2013). The role of the Quantity Surveyor can be identified under the following major skill areas:

a. **Economic**: This involves examining value for money and cost efficiency in design. This depends upon investigating and assessing approaches essential for costing, estimating, and valuing, so employers might be prompted and advised.
b. **Legal**: This depends on general information on the law and specialist knowledge and translation of the law of agreement. This is used to convey contract documentation and settle contractual issues, disagreements, and claims.
c. **Technological**: knowledge of the construction procedure, strategies used to develop buildings and other structures, and in-depth information on the construction sector. This gives a reason for making various capacities.
d. **Management**: The capacity of the Quantity Surveyors to perceive the work related to the development and to affect others in the procurement of buildings and structures, together with capacities of regulatory rules.

The Quantity Surveyor carries out the following functions when engaged in construction projects:

i. **Feasibility studies of capital projects**: Experience has shown that before embarking on any construction work, there is a need to prepare the developer's appraisal or feasibility studies. The financial feasibility exercise, which is the financial statement of the costs and returns of a proposed construction, is prepared to establish:
   - The fair price to pay for the site.
   - The probable building costs.
   - The expected or selling price of the complete projects.

   Thus, the Quantity Surveyor must provide capital investment policy advice. The QS advises on cash flow forecasts, financial plans and procurements, value analysis, sensitivity analysis, cost-benefit studies, life-cycle studies, and cost-in-use. Also, the Quantity Surveyor advises on time effects on costs and profitability and the annual budget for construction projects. This enables the client to have foreknowledge about the feasibility and viability of their proposed project (Cartlidge, 2018; Adunfe *et al.*, 2022).
ii. **Cost modelling/preliminary cost advice**: During the conception phase of the construction project, the Quantity Surveyor prepares the cost estimates

and cost budget for the client. This guarantees that the employer knows the approximate cost of the proposed projects. Also, the QS carries out cost planning, monitoring, and control to ensure the employer's budget is not surpassed. The QS has the responsibility of being involved in cost studies and research. This enables him to offer sound and realistic advice on the probable cost of the project (Chandramohan *et al.*, 2020).

iii. **Cost planning**: Cost planning is an expert practice utilised by the QS. This entails breaking down the cost of building components. This is further analysed to provide the design team with detailed information regarding the quality and utility of a building and to produce a design within an overall cost limit. The QS carries out this function during the planning phase of project development. This is achieved by applying cost criteria to the design process. The outcome will be to sustain a sensible and economic connection between cost, quality, and aesthetics (Cartlidge, 2018).

iv. **Preparation of BOQs**: This tender document shows a list of items briefly describing works and quantities. This comprises building and civil engineering works. The preparation of bills of quantities is the most popular function performed by the Quantity Surveyor. The BOQs serve the following useful purposes in construction projects:

   a. The document forms the uniform basis upon which all tenders should submit their quotations, making comparison and selection easier.
   b. It mitigates the risk intrinsic in the tendering process by offering competitors a qualified description of the work required.
   c. Priced BOQ submitted with tenders could deter the taking of cover prices and make it easier to detect when they have been taken.
   d. BOQ establishes an itemised basis for the tender prices upon which interim payment and valuation of claims are based during the post-contract period.
   e. The document forms a good basis for ordering contract planning and control, including material scheduling.
   f. It provides data for a cost analysis for future cost forecasting.

v. **Advising on contractor selection**: The Quantity Surveyor should provide feasible and realistic advice on the contractor's selection, type, conditions of contract, and other contractual matters. These components have high-cost implications and provide the opportunity of awarding the contract to a competent contractor (Owusu-Manu et al., 2014).

vi. **The QS prices BOQs, negotiates, and agrees with a contractor**: Recent construction projects have become extra complex, leaving no platform for guesswork. One of the outcomes is an increase in construction claims. The QS must ensure that the BOQs are well defined and priced to mitigate this occurrence. This would guarantee that the client is receiving value for money. The contractor is paid a reasonable price for the work executed (Seeley, 1997; Cartlidge, 2018).

vii. **The QS values and reports interim or periodic payments**: Payment to the contractor is on a monthly or milestone basis. The Quantity Surveyor values or measures the work progress as agreed regarding the payment method. This is followed by recommendations to the Architect for certified payment.

viii. **The QS evaluates the cost of proposed variations**: When the Architect or Project Manager issues variation orders, the QS measures the proposed variations and ensures that they are within the original budget.

ix. **The QS carries out a financial report to the client**: The Quantity Surveyor responsible for preparing the financial report either quarterly or monthly (Chandramohan *et al.*, 2020). This shows the current position and the likely total cost of the construction project at completion. This is common with complex projects involving millions of naira. This will give the client a clear picture of how the project proceeds financially regarding budget projection. This is a source of early warning to enable the client to prepare for an increased contract sum at the final account stage.

x. **The QS acts as an expert Arbitrator in a dispute relating to construction**: Disputes in the construction industry are inevitable since the client and the contractor have different objectives. The contractor wants to maximise his profit, while the employer wants value for his money. Disputes between parties in the construction, if not put to check, can derail viable projects and might lead to cost and time overruns.

xi. **The QS prepares and settles the final account with contractors**: The preparation of the final account is one of the most important technical aspects of the quantity surveying profession. This involves the reconciliation of the financial implications of decisions made to fulfil contractual commitments as specified in the conditions of the contract. However, the role of the Quantity Surveyor is pertinent. This is because the final account report reveals the total cost of construction and the amount due to the contractor at the practical completion of the project.

xii. **Other principal services of the Quantity Surveyor**: Other principal services carried out by the quantity surveyor are the following:
- Technical auditing
- Assessing replacement values for insurance
- Value management
- Risk management
- Financial analysis
- Property management
- Asset management
- Property condition appraisals
- Facilities management

This code of conduct implies a bye-law that all practitioners must observe. But before this can be followed, the individual must try to embrace some good rules (principles) to avoid disobeying the laws. They are as follows:

- Responsibility
- Accountability
- Openness and transparency
- Courage to take a stand
- Act within one's limitations

- Credibility
- Faithfulness
- Proficiency
- Clarity of terms
- Dedication to duty
- Integrity: maintaining a good record
- Confidentially
- Competence: Exhibiting the quantities obtained by his training
- Efficiency
- Loyalty
- Impartiality: Do not take sides
- Discretion
- Honesty
- Humanity
- Communication with related disciplines based upon the duty of care by a professional consultant to the public
- Objectiveness

The Nigerian Institute of Quantity Surveyors (1998) new rules of conduct represent the vital minimum structure within which members offer their services. The rules are consistent with national efforts to sustain transparency and accountability in the development process. The rules have six main sections: professional obligations; disciplinary powers and procedures; reinstatement provision; designations; failure to pay money due/to furnish documents to the Institute; and services and notices.

Preamble

1. The constitution, bye-laws, and code of professional conduct, as well as the general laws of Nigeria, govern members of the Nigerian Institute of Quantity Surveyors (hereinafter referred to as the 'Institute'). Decree No. 31 of 1986, which set up the Quantity Surveyors Registration Board of Nigeria (hereafter referred to as the "Board"), regulates the quantity surveying profession.
2. The aim is to regulate the level of discipline, skill, and behaviour of the members of the Institute and the registration board in pursuance of the practice of quantity surveying in Nigeria (NIQS, 1998).
3. The institute's rules of conduct aim to set out and authorise an unmistakable and intelligent system of moral standards. This will advance the obligation of individuals to release their expert commitments to their clients and bosses. This should be done in the interest of the public. To this end, members are expected to embrace good principles and standards, as previously stated, to achieve this goal.
4. The code contains six main parts, with inter-related clauses, which will be of universal application and deal with specific injunctions. Members are expected to be acquainted with all the provisions and use these rules and standards in their dealings with fellow professionals and the public.

5. Failure to adhere to the general standard of conduct contained in this code will be adjudged unsuited to the status of a Quantity Surveyor. This shall attract disciplinary measures, such as reprimand, suspension, or expulsion, as stipulated in the constitution and bye-laws of the Institute and the provision of Decree No. 31 setting up the regulatory board.
6. Principals and partners of practicing organisations shall be held accountable jointly and severally for any contravention of this code committed by any member of their employee in the performance of their official assignment, always provided (it is established beyond reasonable doubt) that the member of staff did not commit a tortuous act or twist the good intention of the partnership or acts on his/her selfish interest.
7. Where appropriate, the rules of conduct contained in this code have been supplemented with commentary to guide standards of good practice beyond the wording of the rule itself. This gives guidance on the interpretation and application of the rules. The commentary, which is of general application unless otherwise stated, is not exhaustive or intended to have legal effect. It is intended to be of persuasive moral value and to notify members of the application and scope of the rules of conduct (NIQS, 1998).

Section 1

Professional Obligations

1. A member who has been commissioned to perform quantity surveying duties or part of the duty must carry out the work to the satisfaction of his employer. This should be done within the rules and regulations approved by the Institute (NIQS, 1998).
2. A registered practitioner's name shall be included in the annual publication of the board on the condition that the practitioner remains a member of the Institute.
3. All quantity surveying documents produced by registered practitioners must be authenticated with the board's personalised official rubber stamps/seals. Every practitioner must validate all their work with their stamp or seal (NIQS, 1998).

Rule 1: Conduct Unbecoming

1. "Honesty, probity, and professional propriety" shall be the watchwords for every corporate member of the Institute.

General Practice Rules

a. Practitioners shall have their service commission evidenced in writing. This should include the conditions of engagement in line with the Institute's current published conditions of engagement and consultancy services agreement or free computations.

b. A practitioner shall not obtain a commission using irregular methods. Any act or permission granted for carrying out any act(s). Or being involved in any act calculated to attract business irregularly to his practice shall be regarded as touting or soliciting.
c. A practitioner, on being engaged or instructed to proceed with professional work upon which another practitioner was previously engaged, shall notify the facts in writing to the original practitioner. He shall satisfy himself that there are no outstandings between the client and the original practitioner before he proceeds to undertake such works.
d. Any attempt by a practitioner to take over the professional work on an on-going project without reference to the practitioner known to be acting for the client shall be treated as a contravention of the rules of the Institute and irregular conduct by the new practitioner.
e. If asked by the client, a practitioner shall be free to give an independent opinion or sanction the work of another practitioner if they do not infringe any provisions of these rules of conduct. Also, their role does not exceed that of giving expert advice or independent assessment.

Rule 2: Conduct of Professional Activities or Business

Advertisement and Publicity

2.1.1 A member is permitted to advertise the professional services offered by his firm or company to the public, subject to the following:

2.1.2 "Advertisement" in this code means any published material or article of any kind whatsoever issued or exhibited by or on the power of a member principally designed to advance the corporate image of the member's business. It includes display and classified advertisements in the press, radio, television, cinema, or elsewhere, posters, features, brochures, cards, leaflets, calendars, blotters, and diaries. Such advertisements shall not contain any comparatives or superlatives.

2.1.3 A member can send articles and scripts on topics related to the construction industry, the professional press, or other information media. The report given in such publications shall be, in substance and presentation, truthful, significant, and neither deceptive nor otherwise discrediting to the profession. The Board would investigate any discredit of the profession by a member following the provisions of Decree No. 31 of 1986 (NIQS, 1998).

2.1.4 Members may advertise salaried employment for staff, giving details of experience possessed or qualifications and experience required and indications of the types of work available, provided that the name of the advertiser may only be shown if such an advertisement is in quantity surveying publications.

2.1.5 Members can advertise their names, addresses, and services, including notices of new or formation of or changes in partnership/company details, by post to their correspondents, classified adverts, and professional press.

2.1.6 Advertisements in newspapers by practicing professional firms must not carry remuneration details.

2.1.7 Members shall not behave or act in any way, publish any matter, cause any matter to be published or distributed, purporting to be or which could be construed as an official action, statement, or publication of the Institute without specific approval of the council. No registered member must permit any reference to them or their firm/company to be shown in any commercial of manufactured products within the construction industry.

2.2.0 Brochures

2.2.1 A member may create a brochure or other materials depicting their firm/company and the services offered. This should only be sent to prospective clients or persons requesting it.

2.2.2 The content of the brochure or leaflet shall:

a. Be truthful and in no way sycophantic of the firm's/company's accomplishments.
b. Avert any explicit solicitation of directives, communicated or inferred similarity with other organisations, or explicit inducements to potential employers.
c. Be intended to ensure that the Institute and the quantity surveying profession are treated disrespectfully. Also, public assurance in the services given by members is not undermined.

The design and layout of the brochure or other materials are left to the practitioner's discretion and should contain the following:

a. The name, address(es), telephone/telefax/telex numbers, and e-mail address(es) of the organisation
b. Truthful report regarding the organisation
c. A list of partners with brief biographical background information confined to matters directly significant to the quantity surveying profession
d. Truthful report regarding the profession and the range of services undertaken by the firm
e. A report that the Institute's code of conduct binds quantity surveying practitioners and that engagement fees are based on the Institute's scale of fees approved
f. A list of projects completed by the firm
g. A list of clients to whom reference may be made and who should have agreed on prior authorisation for using their names

2.3 Letters and Published Articles, Public Speaking and Lectures, Radio and Television Appearance

2.3.1 A practitioner involved in the above-published activities must strictly observe the standards of professional good manners, which require humility concerning personal attainment and achievements and courtesy towards colleagues.

2.3.2 When publicity is not related to quantity surveying, a member may be identified by their name concerning professional designatory letters only without undue publicity of their professional practice.

2.3.3 Should the subject matter being published, broadcast, or circulated for the general public's consumption be relevant to the profession or the built environment, the originating member may state their professional qualifications in addition to their name. He may also state whether he is in full-time employment in the public sector or is engaged in private, professional practice (without their firm).

2.3.4 A member contributing to a professional journal or publishing a textbook concerning the profession or speaking at a course, conference, or meeting with an audience of construction professionals or those studying to become such professionals may disclose their name, their firm or employer, and the position he holds.

2.3.5 For this section, a member must maintain the confidentiality of information passed by their client(s) as privileged information that shall not be for the consumption of third-parties or the public (without express request to do so).

Rule 3: Conflicts of Interests, Impartiality, and Independence

Definition: For Rules 3.1 and 3.3, a person is a "corporate," or a member of such a person is in the member's firm or company, or a partner of his in the case of a consultant firm, or a co-director in the case of a company. Under no circumstances will they allow a third party to interfere with their independent professional service (NIQS, 1998).

Conflict between the Member's and Client's Interests

3.1 Where a contention emerges or may emerge between his advantages or those of any of his associates as defined in the rules and the interests of an employer, a member shall:

1. Disclose to the employer, at the quickest opportunity, the probability of the contention. This should include conditions encompassing it and other important certainties.
2. Inform the employer that neither he nor his firm or organisation can act or keep on representing him except if mentioned to do so, having first prompted the employer to acquire independent professional advice and
3. Promptly affirm to the employer the above position in writing soon.

Conflict Between the Interests of Clients

3.2 Where a contest emerges or may emerge between the interests of employers of a firm or organisation carrying on practice as Quantity Surveyors, of which he is a partner or director, a member shall:

1. Disclose to every employer the probability of the contention. Also, state the conditions encompassing it and some other pertinent realities.
2. Inform every employer that neither he nor his firm or organisation can act or keep representing him except if mentioned to do so. Also, advise every employer first to obtain independent professional advice.
3. Confirm to every employer the situation in writing.

Group Arrangement

3.3 Where the member's organisation is part of a group of organisations, it will not be viewed as a conflict of intrigue if one organisation acts for one employer and another represents the other employer with disparate interests, so long as:
 1. There are no principals or workers in like amid the organization.
 2. There is no immediate or aberrant benefit amid the organisations.

Public Office Holder

3.4 Where a member holds a public position in any field, which may incite a dispute with the interest of any employer of the member, then details of that position and his employer connection must be known to all parties.

Transparency of fees and Benefits

3.5 Every member shall unveil to their employer prompt and in writing the nature and, where known, the premise or amount of any fee, commission, or other advantages (beyond their agreed fee or commission) that they stand to pick up because of their engagement by the employer. Where the disclosure is verbal, the member should verify the divulgence instantly in writing.

Abuse of Appointment

3.6 No member who is directed by an employer to bargain with a third party will specify that he/she will be held by that third party in any capacity except if:
 1. They have their employer's express consent to do so, and this is affirmed in hard copy.
 2. They have counselled the third party appropriately and in writing to receive independent professional guidance.

3.7 No member must, without first advising the third party to obtain independent advice, recommend a transaction concerning landed property, property, or construction, or give preference to a third party solely or partly in anticipation of the procuration of future fee income or other economic advantages, whether directly or indirectly for himself or the organisation.

Duties to Third Parties

3.8 Where a member is to be or has been selected to represent an employer in an agreement in which they owe an obligation of good faith to a third party, he should inform accordingly and in writing the third party any intrigue which they have because of his engagement which goes beyond his normal fee or agreement entered.

Rule 4: Firms/Companies

Name and Style of Practice

4.1.1 The name and style of practice shall, as far as possible, reflect the personal names of the current or past principal partner.

4.1.2 Any member who operates with a pseudonym practice shall indicate the name and qualifications of the partners on the firm's letterheads and documents.

4.1.3 A member who operates as a limited liability company must make sure that the following conditions are met:

i. All directors and shareholders shall be members of the Institute and registered with the Board.

ii. The members' standard professional obligations and conduct are not reduced in any way because of the limited liability company. Adequate protection must be provided for clients by way of professional indemnity insurance. However, a member's name could be sued along with the company should there be professional negligence.

4.1.4 Practicing firms shall be addressed in the "Registered Quantity Surveyor (RQS) style."

4.1.5 Firms shall have consistent letterheads and should inform the Institute where any changes are made.

4.2.0 Partnership and Company Details

4.2.1 Practicing firms/companies shall submit the outline details of their partnership, memorandum, and article of association to the Institute.

4.2.2 All changes that may occur in partnership or company detail from time to time should be communicated to the Institute within six months of such changes.

4.2.3 A practitioner must not have or take a partner or co-director in their firm/company, or any person who is:

a. Disqualified from membership under the provisions of the constitution and byelaws of the Institute.

b. Disqualified from registration by applying the powers conferred on the Board by Decree No. 31 of 1986.

c. Legally incapacitated because of bankruptcy or having been certified as being of sound mind.
d. Unqualified persons or persons other than Quantity Surveyors.

4.3.0 Registration of Practicing Firms

4.3.1 Any member who wishes to establish a practicing firm shall apply to the Institute and the Board, giving all required details. On being approved, he shall have his firm's name entered into the register of firms offering quantity surveying services and approved for the training of persons aspiring to become Professional Quantity Surveyors.

4.3.2 Renewal of registration shall be dependent on the submission by the firms of documentary evidence of the following:

i. The continued existence of the practice and attainment by the firm of a specified number of CPD units
ii. Any changes in the original details of the partnership/company
iii. Compliance with the current practice regulations issued by the Institute and the Board

4.4.0 Consortia

4.4.1 Members shall be allowed to participate in multi-disciplinary firms. All partners/practicing firms/companies are registerable by various professional regulatory bodies named under the Association of Professional Bodies of Nigeria (APBN) and the like.

4.4.2 The scope of practice of a consortium shall be limited to the professional disciplines of the registered participating partners/practicing firms/companies.

4.4.3 The name of the consortia shall, in all cases, be registered with the Institute.

4.5.0 Engagement in Other Businesses/Occupations

4.5.1 A member may be employed in his capability in any business not related to the profession of quantity surveying, whether as proprietor, principal, partner, manager, superintendent, controller, or employees' salaries, consultant to anybody, corporate or non-corporate, or in any other capability, provided his acquaintance with quantity surveying would not be mentioned as part of or form the centre of the said business (where the business involves the performance of quantity surveying duties, the conduct of the member shall obey the provisions of this code of conduct).

4.5.2 A member engaged in the capacity of a salaried employee in government or a government-owned company parastatal involved in construction-related business shall not qualify to offer consultancy services to the government, private enterprises, or government parastatal for any fee. This rule has been relaxed for members of the academic field.

4.5.3 Where a member is a director of a private company within the construction industry or engages in a construction-related business, or an appointed board member of a government parastatal, he may offer consultancy services to other clients but should declare his business interest in the company or parastatal as required in Rule 3.8 of the rules.

Rule 5: Compulsory Professional Indemnity Insurance Regulations

Definitions (NIQS, 1998):

5.1 For these rules, except where the context otherwise requires:

"Member" means:

a. A member who is held out to the public to be practicing as Quantity Surveyor
   - A sole principal
   - A partner
   - A director
   - A consultant

A firm offering quantity surveying services to the client.

The institute's "Professional Indemnity Policy" shall comply with the directives of the Council of the Institute at any time in force. Gross income shall, for the point of Rule 5, imply every professional fee, remuneration, commission, and revenue of any kind to the extent that these have been gotten from work done in Nigeria. This excludes any sums received for the reimbursement of disbursements, any sum fees by way of VAT, and any income from judicial or other such offices as the council may now and again decide:

"Assigned Risks Pool" means an arrangement approved by the council and established by the Institute to provide temporary insurance facilities for members who cannot comply with the Institute's compulsory professional indemnity insurance rules. This is because they have been debilitated.

"Assigned Risks Pool (ARP) Panel" means the panel appointed by Council of the Institute to administer the assigned risks pool on such terms. The pool provides a temporary insurance facility for members, who, for one reason, failed to comply with the Institute's compulsory insurance scheme. The Institute Council determines this from time to time:

"Listed Insurer" means an insurer who:

a. Is authorised by the Corporate Affairs Commission (CAC) and National Insurance Commission (NAICOM) to countersign liability insurance business or NAICOM to provide professional indemnity insurance business in Nigeria.
b. Decides to present a plan that is no less thorough than the type of the Institute and Board professional indemnity insurance plan in force at the time when the policy of insurance is taken out; and

c. Decides to guarantee for one calendar year the ARP line-slip contract on terms set out in the brochure for listed insurers and
d. The Institute lists it in the ARP.

Rule 6: Continuing Professional Development Rules

6.1 For these rules, according to NIQS (1998):
  a. "Member" means a fellow, corporate member of the Institute who has not permanently retired.
  b. "Continuing Professional Development (CPD)" means the system maintenance, upgrading and enlarging of professional knowledge, understanding and skill and the development of professional and technical duties all through the expert's life (NIQS, 1998);
  c. Refer to "Professional and Regulatory Bodies in Chapter One."

6.2 Refer to "Professional and Regulatory Bodies in Chapter One."

Rule 7: Responsibility of Members for their firms

Please refer to the NIQS code of conduct.

SECTION II (Applying the rules in this section places complementary responsibilities on the council and the Board. The Board's statutory powers are neither negated nor derogated from by the provisions in this section. They are to be properly aligned with the enabling powers of the Board.)

Disciplinary Powers and Procedures

Rule 8: Disciplinary Powers

This section of the rule provides what constitutes a contravention that may lead to disciplinary action against a member. For details, refer to the NIQS code of conduct.

Rule 9: Disciplinary Procedure

This section is a follow-up to the previous rule. For details, refer to the NIQS code of conduct.

Rule 10:1 Constitution: Professional Conduct Panels

For details, refer to the NIQS code of conduct.

Rule 10.2: Constitution: Disciplinary Boards and Appeal Boards

For details, refer to the NIQS code of conduct.

Rule 10.3: Complaints and Allegations: General Procedure: Rights and Powers

For details, refer to the NIQS code of conduct.

Rule 10.4: Penalties: Professional Conduct Panel

For details, refer to the NIQS code of conduct.

Rule 10.5: Functions, Procedure, and Penalties: Disciplinary Board

For details, refer to the NIQS code of conduct.

Rule 10.6: Functions, Procedure, and Penalties: Appeal Board

For details, refer to the NIQS code of conduct.

Rule 10.7: Probationers and Students

10.7.1 The same disciplinary procedures for members shall apply to probationers and students.

Rule 10.8: Delegation

For details, refer to the NIQS code of conduct.

Section IV

Designations

Rule 12

12.1 The designation of members by distinguishing initials or words shall be as follows:

a. Every Fellow shall be permitted to use after his/her name the initials FNIQS (i.e., Fellow of the Nigerian Institute of Quantity Surveyors) and RQS (i.e., Registered Quantity Surveyor).
b. Every Professional Corporate shall be permitted to use after his/her name the initials MNIQS (i.e., Professional Corporate of the Nigerian Institute of Quantity Surveyors) and (Registered Quantity Surveyor).
c. Every Honorary member shall be permitted to use after his/her name the distinguishing letters "HonMembNIQS."

12.2 Subject to the Rules

For details, refer to the NIQS code of conduct.

Section VI

Services of Notices and Documents

Rule 14:
For details, refer to the NIQS code of conduct.
Rule 15:
For details, refer to the NIQS code of conduct.
Rule 16:
For details, refer to the NIQS code of conduct.

Definitions

The NIQS (1998) defined the following key terms:

1. "Member" is the person approved by the Executive Council of the Institute as having passed the Test of Professional Competence (TPC) or other equal and related examination of the Institute and duly elected.

2. "Registered member" is the person registered by the Board as set out in Decree No. 31 of 1986.
3. "Practitioners" are persons with the Institute's Corporate or Fellow membership certificates and are registered with the Board. Such persons may be employed in public or private enterprises or partnerships.
4. "Irregular" denotes action not consistent with good professional conduct.
5. "Touting" means using undue influence to procure services or commissions from a practitioner.
6. "Solicitation" means undue requests from clients to procure services or commissions.
7. "Ostentatious" means an unguarded display of a practitioner's practice name to gain admiration or envy.
8. "Election" means a letter from the General Secretary of the Institute confirming the practitioners' membership in the Institute. The letter shall contain the Oath of Good Professional Conduct in the Practitioner's Career and the failure of which the Council and Board may take appropriate steps to protect the Quantity Surveying Profession.
9. "Regulate" is to maintain and improve the high standard of the Quantity Surveying Profession.
10. "Council" means the National Executive Council of the Institute.
11. "Practice Name" is the registered name with the Corporate Affairs Commission in Abuja.
12. "Institute" means The Nigerian Institute of Quantity Surveyors.
13. "Member" means a corporate member of the Institute, namely, Fellows and Corporate Members.
14. "Documents" include all forms of reports, cost engineering documents, bills of quantities, feasibility studies, cost evaluations, and the like produced by the Quantity Surveyors for the client(s).
15. "Board" means the Quantity Surveyors Registration Board of Nigeria, established by Decree No. 31 of 1986.
16. "Professional Press" means publications, journals, and newsletters issued under the sanction of professional bodies like "The Quantity Surveyor."

3.3 MEMBERSHIP, EXAMINATIONS, REQUIREMENTS, AND PROCEDURE

3.3.1 Membership

Membership Grades:

- Fellow
- Corporate
- Honorary Fellow
- Probationer
- Technician

- Student
- Corporate Affiliate Member

3.3.2 Fellow

A prospective fellow must be a financial member of the Institute who has fulfilled the first two conditions below and a minimum score of 80% from the other criteria as highlighted by NIQS (1998), namely:

- They shall have been elected as a Corporate Member for at least ten years.
- They shall have held a responsible position for not less than ten years and have achieved 45 years.
- They shall have partaken effectively in the affairs of the Institute at the national and state chapter levels.
- They shall have presented at least three papers at seminars/workshops/conferences organised by the Institute at the national and state chapter levels, or written five articles on matters involving the construction industry in the Institute Journal, National Newspaper or Local Newspaper.
- They shall have given philanthropic gestures or services continuously to the Institute.
- They must not have been declared culpable or bankrupt by any courtroom or previously suspended or expelled for misconduct by the Institute.
- They must show proof that his membership is financially up-to-date.
- They must have accrued a minimum of 300 CPD points after the election as a corporate member.
- They must have shown proof of active professional practice in the industry.
- All applications shortlisted for upgrading shall be sent to the forum of fellows by the National Executive Council (NEC) for screening following the provisions detailed in this and other sections of the constitution. A minimum of 11 fellows of the forum with seven years of standing nominated by the NEC, including the chairman, shall screen every one of the applications and forward their suggestions to the NEC.
- The Fellow application form can be downloaded from the Institute website.

3.3.3 Member (Corporate Member)

- Members are elected to this membership status, having fulfilled the minimum requirements of the Institute. This includes a member who passed the TPC, diaries and logbook assessments, and the Professional Competence Interview. Any other equivalent international qualification as may be approved by the National Executive Council.
- The National Executive Council may consider a person with professional attainments and distinction attributes and must be at least 50 years old. The person will only need to present himself for the interview to be conducted by the Institute.

- A minimum of 60 CPD points is mandatory to be elected as a corporate member.

3.3.4 Honorary Fellow

An honorary fellow is a person who is not a Quantity Surveyor by occupation but has contributed to the success of the aims and objectives of the Institute through his actions. The national policy committee nominates the person for honorary membership. This is a way of showing appreciation for its contribution to the Institute. Such honorary fellows may only use the designatory letters "Hon. Fellow."

3.3.5 Probationer

A fresh quantity surveying graduate from the university or polytechnic is called a probationer QS. The person can also be involved in quantity surveying duties under the supervision of a Registered Quantity Surveyor. He shall be eligible to sit for the Institute's TPC Examination. The HND holder shall be eligible to sit for the Institute's graduateship Professional Examination before proceeding to the TPC Examination.

3.3.6 Corporate Affiliate Member

Corporate affiliate members of the Institute are organisations or companies that employ Registered and Junior Quantity Surveyors, respectively. Examples are manufacturers, suppliers, banks, practicing firms, construction companies, property development companies, specialist building product companies, etc.

3.4 CONTINUING PROFESSIONAL DEVELOPMENT

In a rapidly changing world of technological development and economic uncertainty, those who offer services to the public must keep abreast of technological changes. This includes constantly updating knowledge and exploiting new technologies to uphold a high standard of skill and integrity in the services they render to the public. Changing professional roles creates the need to update and upgrade the education acquired during formal training continually. This is to imbibe new habits adaptable to the demands of new growth areas of practice specialization because we live in a dynamic world. Clients are increasingly more knowledgeable and demanding specific and efficient service (RICS, 2020). Thus, Quantity Surveyors must project themselves to their clients and address the same marketplace as the allied professions. In that case, we must keep abreast of changes through CPD to guarantee credibility in professional excellence. Quantity Surveyors offer services to the public. Therefore, the services' quality should be of the highest possible standard to retain the public's confidence. The Institute introduced CPD. The main goal is to acquire new skills and knowledge and improve the quality of services offered by Quantity Surveyors.

CPD is a systematic maintenance mechanism that helps to improve and broaden knowledge, skills, and the development of personal qualities necessary for the execution of professional and technical duties in one's working life. CPD training is not a postgraduate study but a learning process designed to continuously upgrade professional persons' knowledge and skills. This will enable the person to remain professionally competent irrespective of inherent changes in professional functions. In this process, no examination is involved. In place of examination, each member is expected to keep a logbook for three years for the yet-to-be corporate members. The Institute has a set of recognised events with an approved number of qualifying hours for each event. If a member participates in any event, the individual enters the total or part of the qualifying hours for that event is entered in the logbook. It is envisaged that each member will devote 20 hours to structured learning each year.

CPD is a watchword among professionals in developed countries where facilities and manpower abound. The relevance of CPD becomes more acute in developing countries such as Nigeria. Consequently, CPD will enable the Quantity Surveyor as a professional, to:

(a) Acquire current and broader post-qualification experience.
(b) Exploit new opportunities and extend their practice.
(c) Offer the public the best and most up-to-date services of his professional experience.
(d) Render services of value for money received from clients.
(e) Promote discipline and public confidence in other professions.
(f) Project his personal quality in the course of his service to the nation and quantity surveying as a profession.
(g) Build a good image for the Institute, thereby inspiring public confidence in the performance of Quantity Surveyors.

The ancient role of the Quantity Surveyor has changed with time from being a measurer as it used to be known. Today, the modern Quantity Surveyor may be called upon to render additional services to his traditional role in any of the following areas: project management, cost control aspects of engineering services, civil engineering, oil rig platform, petrochemical plants, mining engineering, railway engineering, policy formulation in the construction industry, and advice on economic and efficient use of energy through life cycle costing techniques on projects. With the profession's recognition under Decree No. 31 of 1986, the profession cannot afford to fail Nigerians. The professional Quantity Surveyor is expected to be competent and current with a sound and high standard of knowledge in construction matters. This includes offering advice to the public when called upon in any or all the subject matters within his discipline. The usefulness of CPD cannot be overemphasised.

There is a global wind of change in political, social, economic, and technological advancements sweeping today. These changes affect our daily lives and, correspondingly, our profession. The quantity surveying profession cannot be exceptional. Consequently, in anticipation of these changes, we should prepare and keep abreast of any wave of change. Priority must be given to CPD as an essential ingredient in maintaining a high post-qualification competence and integrity standard. Members need to recognise that they offer skilled and specialised services to the public. These services should be updated constantly to meet their clients' changing needs and tastes. As the scope of work and responsibility of the present-day Quantity Surveyor widen, the new emphasis is on professionalism and competence. It has been recognised that some quacks, self-styled, and incompetent Quantity Surveyors still in the profession. However, with Decree No. 31 of 1986, which recognises the Institute and the profession, these bad eggs should be identified and flushed out. Various opportunities are open to the Quantity Surveyor. What is the preparedness of members to show professionalism if called upon to take on challenges? The Nigerian Quantity Surveyor is expected to have the same standard of knowledge and professional competence as their colleagues in the same profession from around the world. Members shall, at any time, be able to demonstrate up-to-date knowledge and ability when the need arises. The time has come for members to look beyond the horizon of the Nigerian market. For example, Quantity Surveying is not widely practiced in some ECOWAS countries, especially in the Francophone Countries. Members should explore services that could be rendered in such countries.

For the Quantity Surveyor to market his skill, he should be up-to-date and be able to demonstrate his professional skills in any aspect of the construction subject matter mentioned. CPD participation is the easiest way the modern professional Quantity Surveyor can fortify his original qualifications and experience with additional up-to-date technical knowledge and its application. The Institute shall issue a practicing certificate to every member that participates in the CPD on the following terms and conditions (NIQS, 1998).

(a) The practice certificate issued to members is subject to renewal every five years.
(b) For a member to be issued a practice certificate, such a member must have satisfied and complied with all the CPD conditions laid out by the Institute.

Members issued with a practicing certificate shall have paid in full their annual subscription at issuance. Members who are in arrears in paying their annual subscription shall be required to pay in full any amount owed to the Institute before issuing the practice certificate.

3.4.1 CPD for All Members

Some people desire to be up-to-date because they want to be competent. Others may do so because of the sheer necessity of being relevant in their work. CPD has been made compulsory for both fellow and corporate members who have been duly elected and are still members of the Institute except those exempted. Before new members are elected, they shall be made to give the undertaking to participate in CPD after their election for as long as they remain members of the Institute, unless they become exempted. To uphold and maintain a uniformly high standard of competence in the profession, the Board ensures that registered Quantity Surveyors make themselves available to participate in workshops/seminars/conferences/training relevant to the quantity surveying profession organised by the Board. The Board also encourages members to fully participate in the Institute's activities. This is to enlighten them on current issues regarding the built environment.

3.5 SUMMARY

This chapter discussed professional ethics, emphasising quantity surveying as a profession, the roles of regulatory bodies, Institute membership examination, requirements and procedures, continuous professional development, and the code of professional conduct of quantity surveying practice. The next chapter gives an in-depth description of tendering procedures and contractual arrangements. Also, it covers international construction, its strengths, threats, and prospects.

REFERENCES

Adunfe, A.A., Moses, E.O., Olalekan, A.H. and Olalekan, S.A. 2022. Assessment of the innovation performance of Nigerian quantity surveying firms. *International Journal of Innovative Research and Development, 11*(3), pp. 11–19.

Ashworth, A., Hogg, K. and Higgs, C. 2013. *Willis's Practice and Procedure for the Quantity Surveyor*, 13th ed. West Sussex: Wiley-Blackwell.

Bowen, P.A., Edwards, P.J. and Cattell, K. 2012. Corruption in the South African construction industry: A thematic analysis of verbatim comments from survey participants. *Construction Management and Economics, 30*(10), pp. 885–901.

Bowen, P.A., Edwards, P.J. and Cattell, K. 2015. Corruption in the South African construction industry: Experiences of clients and construction professionals. *International Journal of Project Organisation and Management, 7*(1), pp. 72–97.

Cartlidge, D. 2018. *New Aspects of Quantity Surveying Practice*, 4th ed. New York: Routledge.

Chandramohan, A., Perera, B.A.K.S. and Dewagoda, K.G. 2020. Diversification of professional quantity surveyors' roles in the construction industry: The skills and competencies required. *International Journal of Construction Management*. https://doi.org/10.1080/15623599.2020.1720058.

Ebekozien, A. 2020. Corrupt acts in the Nigerian construction industry: Is the ruling party fighting corruption? *Journal of Contemporary African Studies, 38*(3), pp. 348–365. https://doi.org/10.1080/02589001.2020.1758304.

Fan, L., Ho, C. and Ng, V. 2001. A study of quantity surveyors' ethical behaviour. *Construction Management and Economics, 19*(1), pp. 19–36.

Nigerian Institute of Quantity Surveyors (NIQS). 1998. *Directory of Members and Quantity Surveying Firms*, 4th ed. Lagos: NIQS.

Owusu-Manu, D.-G., Edwards, D.J., Holt, G.D. and Prince, C. 2014. Industry and higher education integration: A focus on quantity surveying practice. *Industry and Higher Education, 28*(1), pp. 27–37. https://doi.org/10.5367/ihe.2014.0188

RICS. 2018. *Pathway Guide: Quantity Surveying and Construction*. London: RICS. Available at: www.rics.org/globalassets/rics-website/media/qualify/pathway-guides/quantity-surveying-construction-pathway-guide-chartered-rics.pdf.

RICS. 2020. *The Futures Report 2020*. London. available at: www.rics.org/globalassets/rics-website/media/news/news--opinion/rics-future-report-2.pdf.

Seeley, I.H. 1997. *Quantity Surveying Practice*. London: Macmillan.

Yap, J.B.H., Lee, K.Y. and Skitmore, M. 2020. Analysing the causes of corruption in the

Malaysian construction industry. *Journal of Engineering, Design and Technology*, *18*(6), pp. 1823–1847.

Yap, J.B.H., Skitmore, M., Lim, Y.W., Loo, S.-C. and Gray, J. 2022. Assessing the expected current and future competencies of quantity surveyors in the Malaysian built environment. *Engineering, Construction and Architectural Management*, *29*(6), pp. 2415–2436. https://doi.org/10.1108/ECAM-01-2021-0091

4 Tendering Methods and Procurement

4.1 TENDERING PROCEDURES

Choosing an appropriate tendering procedure for a construction project is problematic due to various options and professional advice. Many publications have been issued for several years to assist employers with the tendering procedure. For more than three decades, the National Joint Consultative Committee for Building (NJCC) distributed many respected codes of procedure regarding tendering procedures. The Construction Industry Board (CIB) created its tendering regulations and guidelines when NJCC was disbanded. It is important to state that despite these current rules and regulations, employer organisations still hold fast to NJCC codes of procedure. Likewise, when the Joint Constructions Tribunal (JCT) created its procedure note on tendering, it recognised the consideration of significant material delivered by the NJCC. Construction contractors to be considered for work under consideration are expected to provide information about their technical and financial capabilities. The NJCC has penned the standard form of tendering questionnaire—a private version. This enables the contractors to get responses to significant inquiries ahead of time. The questions mainly address projects that have been conducted in the past few years. Also, the building industry utilised the Code of Procedure for Single Stage Selective Tendering (CPSSST) for many years. The NJCC authors the CPSSST. It was supplanted in 1997 with the CIB code of procedure for selecting main contractors. This method pursues numerous stages or pre-selection and tendering stage actions. Their victory depends on complete designs before tenderers are requested. The following points demonstrate the coverage of the codes:

a. **Preliminary inquiry**: Contractors are allowed to choose whether they wish to tender by receiving an introductory enquiry letter.
b. **Number of tenderers**: The suggested number of tenderers is an upper limit of six.
c. **Tender documents**: The document's objective is to ensure equal opportunities for the tenderers besides pricing/rates.
d. **Time for tendering**: This ought to be at least one month. Additional time may be required, depending on the project's complexity.
e. **Qualified tenderers**: Tenderers ought to do whatever it takes not to fluctuate the basis of their tenders' qualifications. Questions or unsatisfactory agreement conditions should be raised ten working days before tenders are due.
f. **Withdrawal of tenders**: A tender might be acknowledged if it remains open.
g. **Assessing tenders**: They should be opened at the earliest opportunity. Priced bills might be submitted in a different envelope by the contractors.

DOI: 10.1201/9781032661391-4

h. **Examination and adjustment of priced bills**: The Project Quantity Surveyor is responsible for treating the information in the tender documents as classified and reporting mistakes, if any, in computation to the Architect and client. The contractor has the privilege to acknowledge the error and withdraw the tender for correction. A revised tender can be submitted.
i. **Negotiated reduction of tender**: The code of procedure acknowledges the need to search for savings in the cost of a project where the tender surpasses the client's financial limit. This can be accomplished by negotiating with the least tender or the next least if negotiations fail.

4.2 ABUSE OF TENDERING PROCEDURES

The concept behind NJCC Codes and Practices Notes is to convince the stakeholders in tendering to employ a fair and proficient approach. This is the best and most expert strategy being used today. However, this main objective is being threatened because of interests and the absence of time. Some of the difficulties are:

(i) **Large tender lists**: Many practitioners in today's construction industry have widely criticised open competition. Most government establishments at all levels still patronise this approach. However, with the high cost of tendering in mind, many reputable contractors are unwilling to participate in open tendering. One of the potential reasons is that the government agency receives tenders from more than 35 construction contractors.
(ii) **Short, tender periods**: Three factors should determine tendering time: the size of the construction project, the complexity of the construction project, and the standard of the documents. In practice, the design and tender documentation are often late, with employers wanting to make a start on-site quickly, thus eroding the time available for the estimate.
(iii) **Tender documentation**: It is normal in an ordinary system that the Quantity Surveyor ought to get enough drawings to comprehend the nature and extent of the works. Information ought to be provided regarding any restrictions which might affect the contractor's choice of method. Also, the site examination report should be sent to each contractor to avoid problems that might arise during their site investigation. This is unmistakably a tremendous misuse of clearly an enormous waste of effort and a further burden on the already considerable tendering costs, and can be avoided by providing the completed drawings to the construction contractors.
(iv) **Requesting for tenders when the construction project is not likely to proceed**: There is a convention in the construction industry for estimates to be given without charge to the employer. Research has shown that it costs about 0.25% of the tender price to prepare a bid for a traditional lump-sum contract and as much as 2% for the design and build form of the contract. Records have shown that contractors accept this financial risk. This area can be further investigated.
(v) **Qualified tenders and alternative bids**: The tenderer should present his bid without adding conditions to his offer. The contractors should consider

the terms of their offers. There are cases where full compliance with the employer's directives becomes a hindrance. The construction contractor might produce improvements to the design or see an approach for completing quicker. This is often calculated via an alternative price. This new offer is consistent with the earliest brief. Employers can consider alternative tenders.

(vi) **Failure to notify results**: Employers are progressively reluctant to publish figures on the ground that the least tender could endeavour to recuperate the change in value between the tender and the second least, either before the work is awarded or during the project construction.

(vii) **Late receipt of tender documents**: Quantity Surveyors make a valiant effort to manage demands for tenders occasionally at short notice. There are cases where the tender documents are delivered later than promised. This alters the contractor's work programme, and other changes to the tender might be constrained.

4.3 TENDERING METHODS

The decision of a construction contractor to develop a project is a significant issue requiring a cautious approach. An off-base decision may trouble the employer/contractor relationship. The outcome may lead to an insolvent contract. The building contractor is chosen based on competitive tendering in most building contracts. This concurred with the submission that competition is significant (Holt *et al.*, 1996; Brook, 2016). The strategy by which this is accomplished can differ. The next section focuses on the various methods of tendering.

4.3.1 Open Tendering

This type of tendering method is commonly used to select a construction contractor. The client selects a contractor with the appropriate experience and special knowledge to assume the responsibility of construction where he does not want to assume direct responsibility. This selection approach originated with the client's representative. The official, on behalf of the employer, advertises in local newspapers, calling for construction contractors' invitation to apply for tenders in competition to carry out construction projects. The main scope of the work is made known in the advert. Applicants are sometimes required to put down a non-refundable deposit of varying sums of money to collect tender documents. This sum of money indicates a sign of commitment from the tenderers.

Advantages

(i) This allows interested construction contractors to submit a tender.
(ii) This allows the tender list to be made up without prejudice.
(iii) Ensuring good competition and, by extension, giving good value for money.
(iv) Preventing intending construction contractors from forming cabals. The cabal concept, if allowed, will defeat the effort of competition.

Disadvantages

(i) The opening tendering system allows for lengthy tender lists. This involves many construction contractors in pricing, where only one can be awarded the contract.
(ii) Public accountability may be in doubt if the least attractive offer is not accepted. Experience shows that it is sometimes hard to accept the lowest tender price when the contractor's character is questionable.
(iii) Open tendering allows the inexperienced contractor who has misunderstood the complexity of a project to submit the lowest tender. This may cause failure and ultimately lead to bankruptcy.
(iv) There are cases where the lowest offering contractor needs a better management structure. This may affect the performance of the work if the contract is awarded.
(v) In some cases, the contractor decides to submit several claims or reduce the number of materials during construction since he realises that the price is too low. In either circumstance, this will lead to a worsening of relationships between the parties.

4.3.2 Selective Tendering

This is another arm of competitive tendering used in building and civil engineering contracts. This approach requests a list of construction contractors to submit tenders. The list can be drawn up by the employer or the client's professional advisers. An advent could be put in the press for contractors to apply to be considered for the project. This is known as pre-qualification for tender.

Advantages

(i) This ensures that only competent contractors are obliged to tender. One of the outcomes is that it leads to only accepting the lowest tender.
(ii) This tendering method ensures that reasonable profit and overhead are allowed in the rates, ensuring smooth project delivery.
(iii) This tendering method eradicates unnecessary waste of time and effort and reduces the aggregate cost of tendering.

Disadvantages

(i) Care should be taken to guarantee that favouritism does not persuade the inclusion or exclusion of a contracting organisation from the lists.
(ii) Tender prices are always higher than they would have been under open tendering. This is possible because of better administration. This may guarantee the completion date and a higher standard of workmanship.
(iii) This method allows construction contractors who receive documents to submit a high price rather than return the unpriced documents if their name is removed from subsequent tender lists. One of the reasons for this action may be the many projects at hand and wanting to keep the relationship.
(iv) This method increases the more prominent shot of intrigue between organisations, except if the list's composition is changed for each contract.

4.3.3 Negotiated Tendering

This special form of selective tendering encourages a one-name list of tenders. Negotiated contracts are generally entered into for specific reasons. For example, the construction contractor has the unique skill to undertake a specific project requiring a high level of specialised skill within a short possible time frame. If there is no special reason for a negotiated contract, then some other form is likely more suitable for the client. In a negotiated contract, the construction contractor is selected early in the design phase. After the award of the contract, the contractor and the Quantity Surveyor will prepare and price a bill of quantities for the project. Both parties will agree upon the sum in the BOQ. The level of pricing will be laid out in the initial tender document.

Advantages

(i) This tendering method urges the construction contractor to guide the Architect during the plan's development.
(ii) There is a possibility of commencing the work early by ordering materials, prefabricating work, and preparing the work programme on site.

Disadvantages

(i) Evidence has shown that construction projects awarded through this tendering method are usually monopolistic. This results in high prices.
(ii) There is a threat to public accountability as officials involved may likely escalate the contract sum. This is because of the monopolistic nature of the project.

4.3.4 Contract Documentation

The contract documents (Joint Contract Tribunal [JCT]) as cited in Chappell (2013) and Lupton (2019) comprise the following:

- Conditions of contract
- Contract particulars
- Articles of agreement
- Contract bills (or specifications)
- Contract drawings

On civil engineering projects, using the Institution of Civil Engineers (ICE) Conditions of Contract, the following typically represent the contract documents:

- Specification
- Drawings
- Bills of quantities
- Tender
- Written acceptance

- Contract agreement
- Conditions of contract

The complete contract documents for any construction work will include the information highlighted by Ashworth *et al.* (2013).

a) **The work to be performed**: This requires building drawings, including plans, elevations, longitudinals, and cross-sections.
b) **The quality of work required**: The quality and execution of the materials to be utilised and the specifications of workmanship must be communicated to the contractor via the preamble clauses.
c) **The contractual conditions**: A written agreement between the client and the constructor is necessary for everything except for the smallest projects. This will mitigate possible mistakes that are based on assumptions.
d) **The cost of the finished work**: This should be predetermined by an organisation estimate (tender) of cost from the construction contractor. This is not always the case in most projects.
e) **The construction programme**: The period accessible for the construction work on site is pertinent to the employer and the construction contractor. The employer should have an idea of the completion time so that he can plan preparations for the project handover.

4.3.5 Contract Drawings

The contract drawings incorporate the project drawings, from the location plan to details guiding construction activities. The location and site plans show access to the site and the location of the building on site. The floor plans, elevations, and sections are components of the contract drawings. Where these drawings are not given to the construction contractors with the other tendering details, the contractor should be informed where and when he can inspect the site. The inspection creates a platform for informal discussion between the contractor and the design team.

4.3.6 Contract Bills

The bill, or schedule, should include a coordinated list of tasks to be completed. This is together with a description and the quantities of the completed work. The bill may incorporate firm or approximate quantities. This will depend on the comprehensiveness of the drawings and other details from which they were prepared.

4.3.7 Articles of Agreement

This is a component of the contract document and is endorsed by the parties involved. The parties are the client (the building owner) and the contractor (the building contractor). Articles 1 and 2, respectively, summarised the main responsibilities under the contract. Articles 3 to 6 detail contractual and statutory appointments. Articles 7 to 9 detail the dispute resolutions. If the parties make any revisions to the Articles of Agreement or some other portion of the agreement, then both parties should make initial alterations. Executing the agreement as a deed (formerly under seal) under specific conditions might be important or advantageous. This is the situation with the government and other public bodies.

4.3.8 Contract Particulars

The contract particulars for the conditions of the contract comprise major information regarding the project. For instance, the commencement and finish dates, the dates and duration to oblige interim payment, and the correction duration for which the construction contractor is accountable.

4.3.9 Conditions of Contract

This document details the contractor's responsibility to do and finish the project that appeared on the contract drawings. This is described in the bills of quantities to the satisfaction of the Architect (or contract administrator). The document intends to explain the rights and obligations of the various parties in the incident of a disagreement.

4.4 CONTRACT PROCUREMENT

This is the procedure that is utilised to produce construction projects. The implementation of a construction project requires design and construction operations at an on-site location. There are distinctive procurement techniques an employer might need to utilise to get these services. However, procurement methods are becoming flexible and could be used interchangeably to achieve a goal. These procurement strategies give the employer the optimal of several management structures. This includes various contractual arrangements and varying degrees of the employer's risk. The following descriptions provide a broad appreciation of what procurement routes are potentially available to an employer.

4.4.1 Traditional Methods

This is a common and old method of building procurement. This approach conformed to a straightforward system. The conventional methods of building procurement are generally utilised, and their respective distinctive attributes are as depicted below:

4.4.1.1 Based on Bill of Firm Quantities

This is a traditional method that engages the building owner to commission an Architect. The Architect prepares drawings and specification information. Based on the prepared design, the QS prepares the BOQs. Construction contractors are requested to value the bill and submit tenders in a contest for doing the project. The contractor that submits the least tender price is given the contract.

Advantages

(i) Parties have an unambiguous sketch of the degree of their commitment.

(ii) The unit rates in the BOQs offer a reasonable platform for contract variations when necessary.
(iii) The details of the tender sum are expeditiously available.

Disadvantages

(i) It takes a long process and time.
(ii) There is always a challenge in dealing with key variations to the contract.

4.4.1.2 Based on Bill of Approximate Quantities

As the name implies, approximate quantities—the quantities provided in the bill—are estimated and subject to afterward modification. In this type of bill, only the unit rate makes up the contract component. The project may start before the completion of the drawings and the endorsement of the contract by both parties.

Advantages

(i) Construction activities on site may commence in advance.
(ii) The additional cost of preparing firm quantities is circumvented.

Disadvantages

(i) At the tender stage, the BOQ is not the realistic total cost.
(ii) Time-consuming in valuation preparation as construction work must be remeasured.

4.4.1.3 Based on Drawings and Specification

This approach is like the bill of firm quantities. The discrepancy is that no BOQ is provided to the tenderers. The tenderers prepare their quantities from the drawings administered to them. This mechanism is utilised for small, maintenance, and sub-contract works, respectively. Among the attributes is that bidders are given complete working drawings and contract specifications. Also, the practical completion of the working drawings should lead to the endorsement of the agreement.

Advantages

(i) Tender document preparation time is minimised.
(ii) It gives a clearer picture to both parties regarding their obligations when signing the contract.

Disadvantages

(i) Tender sum details are unavailable.
(ii) There are encumbrances associated with the valuation of variations.
(iii) It might be difficult to control the rates for additions.

4.4.1.4 Based on a Schedule of Rates (Measured Contracts or Measurement Contract)

This approach is like a bill of approximate quantities. Tenders are based upon a rate schedule described in (Di)—(Diii). One of the attributes is that it allows for an early start before the contract is endorsed. This is when the drawings are in sketch form.

4.4.1.4.1 Standard Schedule

This is a list of schedules under proper trade headings to confine the items in the construction project. Each item is assigned a unit rate.

Advantage

(i) The assessment of percentage adjustments is easy.
(ii) It is easy for tenderers to get familiar with this type of schedule.

Disadvantages

(i) From the beginning, the contractor and employer are incapable of having a clear picture of their obligations.
(ii) This technique makes it difficult for a Quantity Surveyor to appraise tenders.
(iii) This makes the decision about the most favourable tender not easy.

4.4.1.4.2 Ad hoc Schedule

This type of schedule lists items that are suitable for a specific project. This includes special or unusual terms. The "ad hoc" schedule can be prepared via two methods. The Quantity Surveyor can prepare the schedule, or the rate column may be left empty for the bidder to insert.

Advantages

(i) Parties to the contract understand the scope of work.
(ii) There is a probability that the assessed rates or percentages are more accurate.

Disadvantages

Same as the disadvantages highlighted in standard schedules.

4.4.1.4.3 Bills of Quantities from the Previous Contract

The BOQs utilised are for a similar project to the proposed construction project. This approach, when used in tendering, is called serial tendering.

Advantages

(i) Tender documents are fast to prepare.
(ii) Tenderers are restricted to only the terms associated with the work.

Disadvantages

(i) A precise indication of the respective parties' commitments is not certain.
(ii) There is always a sizeable difference between the contract winner and the real cost of the project.

4.4.1.5 Based on Cost Reimbursements (Prime Cost or Cost Plus)

"Prime cost" signifies the total cost to the construction contractor to buy resources, merchandise, and plant components or hire plants. This includes the engagement of labour to execute the work. The prime cost is extensively recognised as the most uneconomical kind of contract. It should be utilised only when none of the different kinds is appropriate. One of the reasons many use prime-cost contracts is that work on-site can begin in the early stages of drawing preparation. This might be exceptionally imperative to the employer. There might be circumstances where, to the employer, the cost is not a substantial factor compared to time.

Advantages

(i) Tender document preparation duration is reduced.
(ii) It allows an early start to work on site.
(iii) Project activities on site can commence even without a detailed sketch.

Disadvantages

(i) This is to the employer's disadvantage compared to other forms of contract.
(ii) The computation of the total prime cost is tedious.

4.4.1.5.1 Cost Plus Percentage Fee

The construction contractor is paid a charge proportionate to an agreed percentage of the prime costs of labour, materials, and plant utilised in executing the work. The challenge with this approach is that the more ineffective the contractor's operations, the more prominent the misuse of resources, and the higher the charge paid to the contractor will be. For instance, an agreement was expected where the approximate total cost of materials, labour, and plant was N50,000; the contractor's overhead was calculated at 15% and demanded a 5% profit. It was tendered at 20% overhead and the prime cost, expecting that the approximate value of the prime cost demonstrated after the project, to be precise. The total cost would be:

	N	N
Prime cost	50,000	
15% for overheads	7,500	
5% for profit	2,500	10,000
Total cost of contract		60,000

Because of the uneconomic organisation of the contract, the total prime cost was N52,000 for the same project. Hence, the total cost would be:

Total Prime cost	52,000
20% addition for overheads and profit	10,400
Total cost of contract	62,400

4.4.1.5.2 Cost Plus Fixed Fee

This option is not the same as the cost-plus percentage fee. The charge reimbursed to the construction contractor is a sum that does not differ from the total prime cost. This depends on an approximation of the probable total. Using a similar model as in (Ei), the contractor tendered a fixed fee of N10,000 (overheads—N7,500 and profit—N2500). Thus, the financial sketch would be:

	N	N
Total prime cost	55,000	
Overheads	7,500	
Profit	2,500	10,000
Total cost of contract		65,000

4.4.1.5.3 Target Cost

This is one or any of the two past ones, with new variables included. As a motivating force to lessen the total prime cost, the contract rewards the construction contractor only if the total cost is less than an agreed sum (the "target"). They will compensate as a punishment if the total cost exceeds the total amount. The reward and punishment might be a set figure. This is between the total sum and any other concurred percentage. Utilising a similar outline, it is implicit that the fixed fee strategy of disbursement is to be utilised. The agreed sum, based on the probable total cost estimate, is N60,000. At the end of the work, the last payment would be determined this way:

	N
Total prime cost	55,000
Fixed fee	10,000
Total cost of contract	65,000
Deduct: Penalty, being 50% of excess over 60,000	2,500
Amount of final payment	62,500

The charge has been bargained to N7,500. The contractor should be pleased that the concurred sum is a reasonable, probable cost estimate. The Quantity Surveyor should provide a realistic opportunity and help to convince the contractor. The contractor may lessen the prime cost, say to N48,000; the last instalment would be calculated as:

	N
Total prime cost	48,000
Fixed fee	10,000
Total	58,000
Add: bonus, being 50% of saving over N60,000	1,000
Amount of final payment	59,000

The fee increased to N11,000, including N3,500 for profit, and the overheads remained at N7,500. The contractor and client shared the additional expenses or savings in this plan.

4.4.1.5.3.1 Alternative Methods

Since construction activities are dynamic, different approaches to advancing and executing projects have been created with fluctuating degrees of accomplishment. Many owners of construction projects believe that traditional processes are no longer suitable. The reasons are:

(i) The construction cost associated with the mechanism means that large sums of money must be obtained to fund the projects.
(ii) High loan fees imply that the time involved in the traditional processes may bring significant extra construction costs.
(iii) The recent pattern demonstrates that employers are progressively learning about construction matters. They are requesting a better incentive for the value of their money and a prior profit for their investment.
(iv) There is a correlation between digitalisation, innovation installations, and digital construction.

The major alternative methods of building procurement are:

4.4.1.5.3.1.1 Design and Building (Design and Construct or Package Deal)

The mechanism of this method is to mitigate the encumbrances related to the detachment of the design and construction processes. This is responsible for several problems in the built environment. Design and build can defeat this challenge by providing these two separate functions inside a single firm. This firm is the contractor himself. In this scenario, the client approaches the contractor directly against the Architect, which is the traditional method.. With this procurement strategy, the contractor offers the risk for the design elements of the construction project. The employer may engage a QS and Architect on the construction contractor's proposals regarding design, construction strategies, and cash flow. The employer may, in like manner, choose an operator to oversee the work. This is to guarantee that the contractor's recommendations are adhered to.

The terms "package deal" and design and build are interchangeable. Design and build mean a bespoke deal for a one-off project. The package deal is a unique type of design and build project, where the client prefers an appropriate structure, often from a list. The client might be able to view finished structures of comparable design and type constructed elsewhere. This procurement includes industrialised system buildings of concrete, farm buildings, workshop premises, multi-storey office blocks and flats, and low-rise housing. It is still strange for turnkey contracting in some countries, including Nigeria. This has not been used to any large extent. In the Middle East and Far East, the procurement method has recorded notable success. A genuine turnkey contract should contain the necessary components. The components should be from the inception to the occupation of the completed building project. This method gets its title from the turning-the-key idea. When the project is completed, the client can promptly start using the project. Some turnkey contracts require the contractor to find an appropriate site for development.

Advantages

- (i) The contractor is exclusively accountable for any incompetence from the design to completion.
- (ii) The employer has a single individual to oversee.
- (iii) Provided there are no employer changes, the employer knows about this financial obligation from the start of the project.
- (iv) There is healthy communication between the stakeholders. This enhances cooperation and better performance.
- (v) There is an enhanced level of buildability. This may be accomplished because the contractor has a bigger chance of impacting the design.

Disadvantages

- (i) There is an employer's capability to decrease the control of the design.
- (ii) There is high responsibility from the contractor's end preceding the full design.
- (iii) The assessment of the varying design options contained in the contractor's brief may add significant complications to the normal tender review procedure.
- (iv) A much-discounted degree of price information is available to the employer's Quantity Surveyor. This creates important cost-management issues.

4.4.1.5.3.1.2 Management Fee

This is a method of construction procurement. In this approach, the client engages an expert to advise on design and cost matters. Also, a management expert is engaged to advise and supervise the construction project. The project is detailed into individual packages and awarded to sub-contractors. There are two foremost types of management fee methods:

4.4.1.5.3.1.2.1 Management Contracting

In management contracting, the contractor is either employed directly by the employer/client based on their experience or is selected via competition received. The main attribute is that the management contractor is not directly involved in the construction project. The work is split into work packages and contracted to work contractors (sub-contractors). Every one of them signs an agreement with the management contractor. During the construction period, they are paid a fee for the management service and extra service. The major responsibility is offering construction management services for a fee. They must provide and keep all the vital site resources.

Advantages

i. Construction can begin before the detailed sketch is finished. This allows for modification during the construction stage.
ii. Construction professionals are accessible to enhance the design of the project.
iii. To improve experts' contractors coordination via work packages' preparation.
iv. A contractor's understanding of construction costs is utilised to sustain strict budgetary measures.

Disadvantages

i. There is an abuse of variation orders. This is a result of the possibility of making design alterations during the construction period.
ii. It is hard to establish the final cost of the construction project in the early phases. This can be made possible after the last work has been agreed upon and settled.

4.4.1.5.3.1.2.2 Construction Management

This is another variation of the management charge approach, though not well known in this part of Africa. A construction manager is appointed an expert consultant with the authority to examine work on-site and issue directives. The employer has more prominent power over funds during construction. One of the reasons for the power is the agreement with the trade and specialist contractors. They are accountable for organising and arranging the construction work on site. Various contractors do the construction work. Each is engaged in a defined trade package. A trade contractor enters into an agreement with the employer.

Advantages

(i) There is a relationship between the design manager and the construction manager.
(ii) The work is firmly incorporated into the management of the construction project.
(iii) A detailed sketch can proceed in parallel with construction.
(iv) Delay is mitigated because of the privity of the contract agreement (direct approach with each trade contractor).

Disadvantages

(i) There is doubt regarding the construction project's final cost until the last contract's endorsement.

(ii) The employer has more than one consultant and many contractors with whom to bargain, rather than just one main contractor.

4.4.1.5.3.1.3 Government Procurement Routes

The government procurement route is an attempt to enhance construction project procurement within the government sector while simultaneously encouraging efficiency within the construction sector. Recent studies identified three government procurement routes:

- Private finance initiative (PFI)
- Prime contracting
- Design and build

4.4.1.5.3.1.3.1 Private Finance Initiative (PFI)

This is a platform where the public sector contract is used to buy quality services. This is with defined outcomes from the private sector on a long-term premise. This includes maintaining or constructing the essential facilities to benefit from the private sector administration skill (Turner, 2010). Though PFI is one of the favoured procurement routes for government projects, there are confines about its appropriateness. This should be suggested where it provides value for cash and is viewed as appropriate for large construction projects with momentous continuing maintenance obligations. PFI can work in various ways, but build-own-operate-transfer (BOOT) is well known. BOOT suggests contractors be given details of the employer's requests concerning function and maintenance. The bid winner (a consortium of organisations) would oversee the design and construction work. This obligation may run for a period. A payment technique within the PFI agreement allows the PFI contractor to be reimbursed regarding the construction, finance, and maintenance costs. The payments will not commence until the work is completed and is in use. In Nigeria, construction projects such as Lekki Extension Road and International Airport are examples of PFI. If properly managed and parties abide by the signed agreement, PFI would be utilised to provide unique infrastructural facilities across the country and Africa.

4.4.1.5.3.1.3.2 Prime Contracting

This route is uncommon in this part of the continent. This system encourages the utilisation of collaboration and integration into the construction procedure. The prime contractor coordinates and oversees the organisation and management of the activities. It is not a rule that a prime contractor must be a contracting firm. The term "prime contractor" is employed to identify a firm that can provide the above service. The designer, project manager, and contractor can perform the prime contractor role.

4.4.1.5.3.1.4 Project Management

This is not a component of the procurement system but involves general supervision. Project management is the board's planning, controlling, and coordinating of a project from start to finish. This aims to meet the employer's prerequisites so that the construction project will be completed within authorised quality standards, cost, and time (Turner, 2010). Project management activities are getting bigger and progressively more complex. Henceforth, employers must engage the services of an expert to manage the construction projects since the in-house staff skills lack the required experience. Therefore, Quantity Surveyors are eligible to administer project management services because of their training and experience in budgetary and contractual issues, combined with their knowledge of construction procedures.

4.4.1.5.3.1.5 Partnering

Partnering and project management have similar attributes. Partnering is all about bringing trust and cooperation to construction activities. Also, partnering involves stimulating an obligation to a common vision/mission. Others are proffering answers to issues and mitigating adversarial attitudes. An agreement is necessary where partnering is expected to name the partnering to be engaged all over the project. Also, indicate the parties' responsibility for the partnership.

4.4.1.5.3.1.6 Joint Venture

This chapter describes a joint venture as an organisation between at least two organisations. In this context, it would include building, electrical, and mechanical engineering, or other specialist services for tender submission. Each firm has joint and some liabilities for their agreement commitments to the client. This plan for building procurement emerged from the increasing complication of construction projects. It impacts making expert contractors associate with the building contractor. The building contractor under the traditional plan will be a sub-contractor. This does not eliminate the utilisation of subcontractors.

4.5 FACTORS IN THE DECISION PROCESS OF APPROPRIATE PROCUREMENT PATH

The individual project is unique, and new construction approaches lead to the individual project is unique, and new construction approaches lead to increasing demand for choosing the most suitable arrangement (Love *et al.*, 1998). The following variables should be well thought out when choosing a procurement path for a project:

(i) **Construction project size:** This can impact the choice of procurement approach. Smaller-sized schemes proposed for construction projects depend on the conventional procurement method.

(ii) **Cost**: Open tendering is believed to secure the least possible price for a client. It can be argued that competition brings down the price. Price confidence, in this case of pre-measurement or fixed price, might not be the most economical in every case.

(iii) **Time:** Time is constantly an issue to handle in the industry. Many clients want their projects to be completed quickly once they decide to build. In this industry, the design and construction stages are acknowledged to belong and are sometimes delayed. The drawn-out planning processes and approval of the actual design, including the construction stages, are linked to the causes of the delay. Because of this delay, some clients opted for an alternative approach, such as the package deal type. This method can enhance the speed of construction.

(iv) **Accountability:** Every employer needs the affirmation that they have gotten the safest possible procurement approach against their list of objectives. Satisfying the accountability criteria regarding price is difficult because of the absence of competition. Accountability is interwoven with money. Thus, the emphasis is on paying the least cost for the finished project. Showing some clients that paying more for a supposed higher quality or prior finish is worthy. This might be easy, but only in some cases.

(v) **Design:** A design and build contractor is expected to be able to attain an answer which considers buildability. There are cases where the clients, having been furnished with an unacceptable building, must wait eagerly while the consultant and contractor argue about their responsibility (Bolomope *et al.*, 2022). Therefore, the design variables to consider are aesthetics, function, maintenance, buildability, contractor involvement, standard design, design before the build, and design prototypes.

(vi) **Quality assurance:** The quality of buildings relies on an entire scope of inputs: the suitability of the design, a correct choice of fulfilment, proficient working details, adequate supervision, and the ability of the developer. The skills of the workers are also important. This is because the current trend of designs tends to become more complex in their detailing. So, a higher level of skill and experience is required. Quality control, independent assessment, group working, harmonisation, future upkeep, design and detailing sketches, and the character of craftspeople are the quality assurance factors that should be considered (Smith *et al.*, 2004).

(vii) **Organisation:** Studies have shown that the organisation and contractual arrangements are more complex because of the many parties engaged in a building project. It has been suggested that the engagement of a few workers will mitigate organisational challenges. The major organisational variables to be considered are a complication in the arrangement, single obligation, levels of duty, number of individual organisations engaged, and administration lines.

(viii) **Complexity:** Complexity is one of the results of innovative design. This is the utilisation of a new construction method. Construction projects are multifaceted in structure and require exact and thorough contractual arrangements. The nature of complexity, the parties' abilities, and the main aims of the client are some of the organisational points that should be considered.

(ix) **Risk:** Risk is a potential misfortune resulting from the difference between what was anticipated and what happened. Risk is inbuilt in the structure and construction of a building. Risk is not completely monetary. The client aims to move this risk to the consultant or the contractor. The risk moves from the client to other parties engaged in the project might seem to fulfil the responsibility criteria. The degree of risk varies from project to project.

(x) **Finance:** Traditionally, the contractor could be paid in two ways payment via stage or monthly payments. These payments would assist the contractor in balancing the budget regarding salaries, goods, and materials for the project.

4.6 INTERNATIONAL CONSTRUCTION

International construction projects demand complex multi-mechanism analysis. This is because the competition in the market could be fierce (Li *et al.*, 2020). For this book, international construction implies that one firm, resident in one country, performs construction work in another nation. This is the simplest of the definitions. But this description has raised some issues on the global platform. The issues include the identification of nationality: for example, a Nigerian-based company working in the country but owned by a Ghanaian company. Legitimately, it would be viewed as a Ghanaian company working internationally. This mechanism has enhanced International Construction Joint Ventures (ICJV). It is a novel form of strategic alliance used globally for delivering complex engineering construction projects (Shen and Cheung, 2018; Chan *et al.*, 2020; Tetteh *et al.*, 2020). In the case of an international project, the meaning is clear. It is a project that two or more nations will profit from, directly or indirectly. International construction is much more dangerous than domestic construction. The international environment is pretentious by diverse factors that are not part of the domestic conditions. Construction contractors from a few advanced countries dominate the international construction market. American firms had the highest (14) in annual Engineering News-Record surveys during 1990–1999, followed by developed countries such as Germany, Britain, Japan, and France. ICJV could be described as a temporary relationship between at least two construction firms (i.e., different countries of headquarters locations) who agreed to put resources in pursuit of construction projects (Hong and Chan, 2014; Tetteh *et al.*, 2020). Globally, there are successful ICJV projects. This includes the Hong Kong-Zhuhai-Macau Bridge, the Three Gorges Dam in China, and the channel tunnel between the United Kingdom and France (Liang *et al.*, 2019). The developing countries are gradually increasingly embarking on these international projects like their counterparts (developed countries). According to ENR's, as cited in Viswanathan and Jha (2020), of the top 100 international construction firms, 43 are from emerging countries. This includes Korea, India, Turkey, and China.

4.6.1 Strengths of International Construction Companies

The following have been identified as the strengths of international construction companies:

1. **Track record:** This is one of the highest-ranked strengths of a firm relative to international construction. An accomplished firm has either a prepared solution or a less expensive one to a technical issue. This is because it has experienced a related challenge before and has invested in its solution. Track record is important in expert engineering, project management, and large contracts.
2. **Specialist expertise:** This is one of the highest-ranked strengths of the international arena. The major share of the power, industrial/petroleum, hazardous waste, and sewer/waste markets is in the United States of America. Specialist technologies enable less significant firms to carve a niche in the international market by challenging specialist subcontracts or as a

desired consortium partner. The provision of novel technologies is a strategy for getting work in an environment where low-cost pricing may favour a home-based construction company.

3. **Project management capability:** This is one of the strengths of international construction companies. Many researchers have observed the importance of project management capability in international construction. International projects are like complex projects. Multiple ownership, elaborate financial provisions, and different political ideologies are attributes of international construction companies' project management capability. These construction projects are harder to operate than local projects. This is because the risks involved are many and less certain.
4. **International network:** An international marketing set of connections allows an organisation to obtain information on technology, up-coming projects, buyers, intending competitors, and intending co-venturers. This information permits a construction firm to devise a suitable competitive method. Construction organisations fascinated with discovering international markets might engage in an international trade mission. This is to create international connections. This business mission is usually coordinated by trade associations or the government (Viswanathan and Jha, 2020).
5. **Technological advantage:** This is discussed by the scope of the knowledge and professionals utilised in technically sophisticated projects. Technology is regularly one of the successful missiles that enable a construction firm to break through international markets. This is urgently needed in developing countries, such as Nigeria. It will enhance technological advancements regarding infrastructural facilities.
6. **Financial strength:** The stronger a firm's financial position, the better it can complete extensive and sharp strategic plans. It can also take higher risks with the possibility of better yields. It appreciates a higher level of validity and reputation among its clients and suppliers. Also, with the capacity to offer pretty financing bundles to would-be employers, they will choose which foreign contractor will win the bid. A decent balance sheet is the main qualification for securing attractive financing bundles from financial organisations.
7. **Equipment, material, and labour support:** A contractor's abilities concerning qualified personnel, equipment, and plant institute significant factors to be considered in evaluating intending bidders in global construction.

4.6.2 Threats Linked with International Construction

The addition of international markets to a firm's portfolio unavoidably brings along a certain element of risk and indecision. These risks include investment, currency, and commercial risks. Some of the threats identified are as follows:

1. **Loss of key employees:** Losing experienced and gifted staff is the most important threat connected with global construction. A firm's personnel are the most significant resource in a competitive environment and key to its accomplishment.

2. **Shortage of project owner's financial resources:** No shortage of global construction work is needed. There is a lack of capital or capability concerning the owners to balance financial resources against construction needs.
3. **Inflation and currency fluctuations:** Foreign market contracts involve the continuous exchange of funds across national limits. Money variance is enthusiasm for construction organisations operating in third-world countries from industrialised nations. A potential ramification of declining local cash in a developing nation is the expense of imported materials, plants, and equipment. This will rise, thereby diminishing the amount of anticipated profit. The results of inflation in the host country also soberly affect the cost of a construction project.
4. **Interest rate increases:** International construction companies searching for financing must battle among major financing sources: countertrade, investment bankers, commercial banks, international financial institutions, and national export credit agencies. Interest rates are germane in the accomplishment of international construction firms.
5. **Foreign competitors in the host country:** This is a potential danger for new contestants in the market. In the changed circumstances, foreign contractors face a mind-boggling circumstance in which achievement only partially relies upon price competition.
6. **Cultural differences:** During business across national limits, there is communication between individuals and their organisations. This involves different cultural environments. Difficulties experienced in foreign projects often find their genesis in the contrasts between the cultures of the international contractor and the other stakeholders in the host nation (Hwang *et al.*, 2017).
7. **Bribery in the host country:** Bribery and corruption are unethical words in the global market. Attempting to be ethical and principled is always challenging in an environment where unethical practices are illegally accepted as part of the system.
8. **Conflicts among ICJV entities:** Establishing a conflict-free ICJV relationship is difficult, but conflict can be mitigated if well planned. Han *et al.* (2019) asserted that the goal incongruences among ICJV stakeholders might begin with the difference in the core benefits anticipated by each organisation.
9. **Policy and political issues:** The government interference by one ICJV partner to conduct the business regulations would significantly impact the ICJV's success. Extreme regulations and policies would endanger ICJV progress (Tetteh *et al.*, 2020).
10. **Legal issues:** In many instances, there are uncertainties surrounding the contractual agreements of ICJV, which might deter its success (Li *et al.*, 2020). This is compounded by different countries' contractual clauses, codes, and regulations.

4.6.3 International Construction in Future

Many assessments show that more construction is needed worldwide to meet the fundamental needs of the world's developing population. This will help enhance and upgrade the quality of life. Industrial and other business offices will be increasingly unpredictable owing to technological advancement. Information technology has influenced the nature and internal environment of productive facilities. The new advancements will facilitate efficiency and productivity drives at all stages of construction procedures. Expected future accomplishments in global construction include leadership, social acceptability, cost-effectiveness, novelty, and firm effectiveness. The prospects of future international construction firms are a competitive advantage for the global industry. This includes the ability to procure worldwide, experience and reputation commitment to carry risks, technical skill, decreased project timescales, and project finance. Others are the ability to adopt company structures to work in multi-firm, multi-cultural, and multi-profession networks; the capability to provide project funding; the capability to form partnerships or alliances with organisations with skills in construction or other areas such as finance, design and operation, and management; and the re-use of information and political backing in corporate infrastructure.

4.7 SUMMARY

This chapter discussed types of tendering procedures and contractual arrangements, including merits and demerits. Also covered in this chapter were international construction, its strengths, threats, and prospects. The next chapter gives an in-depth description of contract administration, including measurement of variation and interim valuation, final accounts, daywork and variation account preparation, fluctuation, preparation of prime cost, and provisional sums accounts.

REFERENCES

Ashworth, A., Hogg, K. and Higgs, C. 2013. *Willis's Practice and Procedure for the Quantity Surveyor*, 13th ed. West Sussex: Wiley-Blackwell.

Bolomope, M., Amidu, A.R., Ajayi, S. and Javed, A. 2022. Decision-making framework for construction clients in selecting appropriate procurement route. *Buildings*, *12*(12), 2192.

Brook, M. 2016. *Estimating and Tendering for Construction Work*. London: Taylor & Francis.

Chan, A.P., Tetteh, M.O. and Nani, G. 2020. Drivers for international construction joint ventures adoption: A systematic literature review. *International Journal of Construction Management*, *5*, pp. 1–13.

Chappell, D. 2013. *Understanding JCT Standard Building Contracts*. London: Routledge.

Han, L., Zhang, S., Ma, P. and Gao, Y. 2019. Management control in international joint ventures in the infrastructure sector. *Journal of Management in Engineering*, *35*(1), p. 04018051.

Holt, G.D., Olomolaiye, P.O. and Harris, F.C. 1996. Tendering procedures, contractual arrangements and Latham: The contractors' view. *Engineering, Construction and Architectural Management*, *3*(1/2), pp. 97–115. https://doi.org/10.1108/eb021025

Hong, Y. and Chan, D.W. 2014. Research trend of joint ventures in construction: A two-decade taxonomic review. *Journal of Facilities Management*, *12*(2), pp. 118–141.

Hwang, B.G., Zhao, X. and Chin, E.W.Y. 2017. International construction joint ventures between Singapore and developing countries: Risk assessment and allocation preferences. *Engineering, Construction and Architectural Management*, *24*(2), pp. 209–228.

Li, G., Zhang, G., Chen, C. and Martek, I. 2020. Empirical bid or no bid decision process in international construction projects: Structural equation modeling framework. *Journal of Construction Engineering and Management*, *146*(6), 04020050.

Liang, R., Zhang, J., Wu, C., Sheng, Z. and Wang, X. 2019. Joint-venture contractor selection using competitive and collaborative criteria with uncertainty. *Journal of Construction Engineering and Management, 145*(2), p. 04018123.

Love, P.E., Skitmore, M. and Earl, G. 1998. Selecting a suitable procurement method for a building project. *Construction Management & Economics*, *16*(2), pp. 221–233.

Lupton, S. 2019. *Which Contract?: Choosing the Appropriate Building Contract*. London: RIBA Publishing.

Shen, L. and Cheung, S.O. 2018. How forming joint ventures may affect market concentration in construction industry? *International Journal of Construction Management*, *18*(2), pp. 151–162.

Smith, J., Zheng, B., Love, P.E.D. and Edwards, D.J. 2004. Procurement of construction facilities in Guangdong Province, China: Factors influencing the choice of procurement method. *Facilities*, *22*(5/6), pp. 141–148. https://doi.org/10.1108/02632770410540351

Tetteh, M.O., Chan, A.P., Darko, A. and Nani, G. 2020. Factors affecting international construction joint ventures: A systematic literature review. *International Journal of Construction Management*, 1–17. https://doi.org/10.1080/15623599.2020.1850203

Turner, F.D. 2010. *Quantity Surveying Practice and Administration*. London: George Godwin Limited.

Viswanathan, S.K. and Jha, K.N. 2020. Critical risk factors in international construction projects: An Indian perspective. *Engineering, Construction and Architectural Management*, *27*(5), pp. 1169–1190. https://doi.org/10.1108/ECAM-04-2019-0220

5 Contract Administration

5.1 MEASUREMENT OF VARIATIONS AND INTERIM VALUATION

5.1.1 Measurement of Variations

The word "variation" cannot be overemphasised in the built environment. After an agreement has been endorsed, making changes by the parties might be difficult. But, given the idea of the construction procedure and the entirety of its uncertainty and risk characteristics, most standard construction contracts incorporate the provision for variations or modifications to the works. The term variation is defined in Clause 11.2 of the Federal Ministry of Works (FMW) Lump Sum Contract as "the alteration or modification of the design, quality or quantity of the works as shown in the contract documents." The non-appearance of such a provision within the agreement conditions would require another new agreement to be drafted if variations emerged. The side effect of such a provision is that it frequently enables the Architect to defer making some decisions almost until the last conceivable minute (Mustaffa *et al.*, 2023). Variations may emerge in any associated circumstances (references are to the FMW Form): The FMW 11.1 makes such variation matter. The Architect or supervising officer may issue instructions or sanctions regarding the design or specification in writing variation to the works carried out, otherwise than according to his instructions.

1. When an inconsistency is revealed between any statutory requirement and any agreement documents, FMW 4.1 authorises the Architect to issue instructions upon receipt by him of a written notice by the contractors specifying the divergence. Clause 4.1 further stipulates that "if within seven (7) days of having given the said composed notice, the contractor does not receive any instructions regarding the matter therein specified, he shall proceed with the work affirming to the act of parliament, instrument, rule, order, regulation or bye-law in question and any variation thereby necessitated shall be deemed to be a variation required by the Architect."
2. When inconsistency is uncovered among two or more of the agreement documents, Clause 1.2 requires the contractor to provide written notice to the Architect/supervising officer specifying the discrepancies or divergences between those documents. The Architect shall issue instructions as demanded.
3. When a mistake or omission from the contract bill is found, Clauses 11.1 and 11.2 provide that any error in the description or quantities or any omission of items from the agreement bill shall not vitiate the contract but be remedied and considered a variation.

The biggest number of variations emerges under (1). In cases (1), (2), and (3), the Architect is required to give guidelines, although not significant in case (4). Every

DOI: 10.1201/9781032661391-5

instruction must be documented in writing. Variation orders, being instructions from the Architect, must be in written form. Clause 2.3 refers to oral instructions given by the Architects or instructions given by the clerk of works on behalf of the Architect before they take effect. Clause 11.6 aims to ensure that the contractor is refunded for any loss or expense caused by the regular progress of the work being materially affected because of compliance with a guideline of the Architect requiring a variation. Variation orders are used to:

i. Effect desired changes in the quality, design, or specification.
ii. Resolve discrepancies between statutory requirements and any of the agreement documents.
iii. Correct errors, wrong descriptions, incorrect quantities, and omissions of items in the delivery contract bills.

5.1.2 Basic Rules for Valuation of Variations

Variations require the re-measurement of the construction works on the project site or from contract drawings. The main contractor is privileged to be present when such measurements occur. Significantly, such measurements are concurred upon by both parties (Mustaffa *et al.*, 2023). The various contract forms clarify and explain how such variations can be priced. The description below presents an illustration of the variation in valuations. This illustration takes into consideration the set of principles in the various contracts. This is in line with Clause 11.4 and is as follows:

1. The variations shall be measured and valued by the Quantity Surveyor. The Quantity Surveyor shall offer the contractor the opportunity to be available and to accept such notes and measurements as he may need, except the Architect acknowledges an estimate made by the contractor as a lump sum.
2. Where the extra or substituted work is identical in character, conditions, and quantity to items in the agreement bills, then bill rates or prices are used to value the variation.
3. The extra or substituted work is the equivalent, yet it is executed under various conditions or results in noteworthy changes in quantity. The bill rates, or prices, are used to value the variation. This is known as pro-rata rates.
4. When it is unfeasible by the methodologies of measurement or price build-up to exemplify the cost of the work executed, then the work can, by agreement of the Quantity Surveyor, be permitted at day work rates at the price ruling at the date the work is done (except if, obviously, generally accommodated in the specific contract documents).
5. Where the approximate quantity is not an accurate forecast of the quantity of work, the rate or price quoted for the approximate quantity will be utilised as a premise, and allowance for such contrasts will be utilised.
6. Omissions are regarded at bill rates, except if the rest of the quantities are subsequently changed.

5.1.3 Variation Accounts

The procedure involved in variation accounts starts with taking measurements, where the dimensions must be worked up into variation accounts. This follows the FMW form. The variation accounts should be completed and priced. A copy of the priced variation accounts should be forwarded to the contractor nine (9) months after practical completion. Practically speaking, this may not generally be accomplished. The QS must complete the variation account by the expiration of the period expressed in the agreement. This may result from breaching the agreement on the client's part if the Quantity Surveyor fails to do so.

The form in which variation accounts are drawn up changes between one organisation of Quantity Surveyor and another. A few organisations bill omissions and additions under every Architect's Instruction (AI) independently. Others merge all of the omissions and the additions together, with every group being arranged under proper work section element headings. The advantage of the past system is that the cost of each variation is evident, together with separate totals of omissions and additions. The main hindrance is that variation accounts are practically sure to form a thicker document because of the repeated items in various variations. One of the reasons is the extra space taken up with separate totals. The subsequent strategy brings about fewer items and a more compact document.

5.1.3.1 A Practical Example of a Variation

An extract relevant to the variation orders four, five, and six from the contract bill of quantities.

Items in the Contract Bill	Unit	Qty	Rate Amount
1. Excavate topsoil average, 150 mm deep.	m^2	80.00	
2 Remove topsoil from the site.	m^3		650.00
3. 300-mm bed of broken hard-core to make-up level.	m^3		6,000.00
4. 75-mm precast concrete paving slab, size 375 mm x 375 mm, laid on hardcore.	m^2		550.00
5. 175 mm × 75 mm precast coping weathered on top, twice throated, fair finished surfaces bedded and joined in cement mortar (1:4).	m	18	1,300.00
6. 250 mm × 75 mm ditto	m	16	1,750.00
7. Plain in-situ concrete (1:3:6—19 mm agg) in beds 75-mm thick laid on hardcore.	m^3	18	25,000.00
8. Reinforced in-situ concrete (1:2:4—19mm agg) in a suspended roof slab 150-mm thick.	m^3	14	30,000.00
9. 16-mm-diameter mild steel straight and bent bar reinforcement to BS4449 in suspended slab	Tonnes	1.27	261,000.00

Items in Variation Order

AI Nos.	Date Issued	Subject
4	7/5/2013	Introduce 75-mm-thick precast concrete paving slabs in the soakaway area of the building (area of soakaway is 4500 m long and 2500 mm wide).
5	29/9/2013	250 mm x 100 mm precast concrete coping in lien of 175 mm × 75 mm.
6	13/12/2013	Concrete (1:2:4–19mm agg) in bed as against concrete (1:3:6–19 mm agg)

AI No: 4

Dimensions	Squaring	Description	Unit	Qty	Rate	Amount
		Date issued 7/5/2013				
		Measured 10/5/2013				
		Description				
		75-mm-thick precast concrete paving slab in the soakaway area of the building				
		Addition				
4.50	11.25	Excavation topsoil to be preserved averages				
2.50 m^2	11.25	150 mm deep.	m^2	11	80.00	880.00
4.50						
2.50		Remove topsoil from site.	m^3	2	650.00	1,300.00
0.15 m^3	1.69					
	1.69					
4.50						
2.50		300-mm-thick broken stone hardcore filling to				
0.30 m^3	3.38	make-up level	m^3	3	6000.00	18,000.00
	3.38					
		75-mm precast concrete paving slab in 375 mm				
		× 375 mm laid on hardcore and jointed in	m^2	11	550.00	6,050.00
4.50		cement mortar (1:4)				
2.50 m^2	11.25					N26,230.00
	11.25	Addition: Carried to Summary				
		Note: Contract bill rates have been used to determine the value since similar work has been set out in the contract bill.				

AI No: 5

	Unit	Qty	Rate	Amount
Date issued 7/5/2013				
Measured 10/5/2013				
Description				
250 mm × 100 mm precast concrete coping in lien of 75 mm × 75 mm	m	18	1,300.00	23,400 N23,400
Area of the building				
Omit				
Page . . . item . . . (as in the contract bill)				
Omission carried to summary				
Add				
250 mm × 100 mm precast concrete coping weathered on top, twice throated, far finished surfaces, bedded and jointed in cement mortar (1:4)	m	18	2,250	40,500 N40,500

Addition: carried to summary

Note: Rates in the contract bill have been used as bases for calculating the new rate that was inserted.

250 × 75 mm coping
= 18750 mm² @ N1750

175 × 75 mm coping
= 13125 mm² @ N1300

Deduct 5625 mm² @ N450

∴250 × 100 mm = 25,000 mm²
175 × 75 mm = 13,125 mm²

Addition area = 11,875 mm²

If 5625 mm² additional area cost N450

∴11,875 mm² additional area will cost

$$\frac{11875\ 450}{5625} = N950$$

New rate = N1300 + N950 = N2250

AI No: 6

Description	Unit	Qty	Rate	Amount
Date issued 13/12/2013				
Measured 16/12/2013				
Description				
Concrete 1:2:4–19 mm Agg in bed as against concrete 1:3:6–19 mm Agg.				
Omit				
Page . . . item . . . (as in the contract bill)	m^3	18	25,000.00	450,000
Omission carried to summary				N450,000
Add				
Reinforced concrete 1:2:4–19 mm Agg in bed poured on hardcore	m^3	18	30,000.00	540,000
Addition: carried to summary				N540,000
Note: The contract bill rate was used as a basis since only the specification of concrete was changed. The value of concrete (1:3:6) was determined and isolated, and the value (1:2:4) was placed with the new rate.				

Summary of Variation Account

Project Title: Four-Bedroom Bungalow
Client: Mr. Monday and Mrs. Joy Asemota
Date of commencement: September 20, 2020
Contract sum: NGN3,171,500.00 kobo
Revised contract sum: NGN3,308,163.25 kobo.

SUMMARY OF VARIATIONS

Items	Nature of Work	Omission	Addition
4.	Substructure	———	26,230 =
5.	Concrete coping	23,400	40,500 =
6.	Concrete bed	450,000 =	540,400 =
		473,000 =	606,703.00k
	Add: water and insurance @ 25% (assumed)	11,835 =	15,168.25k
		485,235 =	621,898.25k
	Net addition: carried to statement		485,235
			136,663.25k
	Statement		
	Contract sum		3,171,500 =
	Net addition for variation		136,663.25k
	Revised/adjusted contract sum		3,308,163.25k

5.1.4 Interim Valuation

The employer/client is expected to make an interim payment to the contractor for an ongoing construction project. This alleviates the contractor's burden of financing the project from start to finish. This is because work may take various months or years to wrap up. The reason for valuations is to enable the contractor to get payment on account. In this way, the contractor is financed during the advancement of the work. Likewise, the construction contractor's expense of borrowing a large sum of money to finance the construction project is mitigated. Therefore, interim valuation is the value of work carried out under a construction contract, including materials and goods delivered to the project site at some defined intervals during the advancement of the works. The Quantity Surveyor prepares an interim valuation. This allowed an interim certificate to be issued to the contractor by the Architect based on the prepared interim valuation received from the Quantity Surveyor. The FMW condition of the contract offers that the Quantity Surveyor should prepare the interim valuation. This should be done at whatever point the Architect believed them to be necessary to ascertain the amount to be expressed as due in an interim certificate.

However, two alternative approaches to the intervals for the certificate are provided by the conditions of the contract:

(i) Valuation shall be at regular intervals, including after practical completion. It is normally a one-month interval or any other agreed-upon term to be stated in the appendix.
(ii) An alternative to regular monthly valuation is to agree on stage payment regarding the contractor payment at defined intervals in the physical progress of the work. That is, valuations can be prepared at convinced defined phases during the construction. Thus, the first valuation may be done when the substructure is completed, the second when frames, the third when the roof and upper floors are completed, etc. The stage payment method is suitable for small jobs, not based on quantities (Oyegoke *et al.*, 2022).

The FMW condition of contracts provides that the contractor must be eligible for payment within the duration of the honouring certificate named in the appendix. The said appendix provides that if a period is not stated, it shall be twenty-eight (28) days from the date of the certificate released. In preparing a valuation, it is normal to calculate the amount of work that has been executed since the start of the project. Also, the value of goods and unfixed materials on site is included in the valuations to arrive at the total value. Note, previous certificates issued, if any, will be subtracted, as will the balancing process for payment. The interim valuation form comprises the project's information section and the payment recommendation build-up column. The two separate segments contain and provide information as follows:

(A) Project information
  (i) Project title
  (ii) Client's name and address
  (iii) Contractor's name and address

(iv) Contract sum
(v) Contract commencement date
(vi) Completion date
(vii) Revised contract sum
(viii) Valuation number
(ix) Date of valuation
(x) Date of issuing valuation
(xi) Retention(percentage and limit-of)

(B) Payment recommendations build up the column.
In this form segment, the determined work values are set out successfully, one after the other, in a column down the line until the determination of net valuation at the bottom of the column. The components of the build-up are:
(i) Value of work completed
- Main contractors
- Nominated sub-contractor
- Nominated suppliers
- Material on site
- Sub-total (value of completed works)

(ii) Deduct retention.
(iii) Fluctuation (sub-total, gross payment due/authorised).
(iv) Add advance payment authorised and paid.
(v) Deduct the authorized advance payment.
(vi) Add vat 5%.
(vii) Deduct the previous payment certified to date (total payments excluding advance payments).
(viii) Sub-total: net valuation recommendation for certification.

5.1.5 Detailed Breakdown of Valuation

A detailed breakdown of the valuation build-up provides greater features of the components of the works and their values as listed under the summary sheet in the valuation form. The respective headings under which the details shall be presented may include the following:

(i) Main contractors work
- Preliminaries
- Various building types or work sections
- External works

(ii) Sub-contractors and suppliers
(iii) Fluctuations and authorized claims
(iv) Re-measured work
(v) Variation
(vi) Materials on site
The aforementioned list shows a typical valuation form. A detailed breakdown of valuation items shown and summarised on the form shall be based

on the contract bill or re-measured works valued at contract rates. This indicates the value of the construction work done to date.

(i) Advance payment.
   - The release of advance payment should be preceded by an agreed-upon schedule of the material and components proposed to be procured for stocking or ordered for advance procurement. The amount for the procurement shall be determined by applying basic material and component prices that are appropriate to the contract rates.
   - Repayment terms (amortisation) should be agreed upon at the advance payment negotiation stage before the formalisation of the contract. Clients should approve advance payment under the contract terms to protect the client's financial interests.
   - Materials and components covered by the advance payment shall not be subject to fluctuation, having been covered by the pre-payment.

(ii) Defective work.
   - It has been variously recommended that a provision be made on the valuation form for defective works. The deduction of completed but defective works should be backed with appropriate instructions, upon which the contract value of the works would be omitted.

(iii) Nominated sub-contractor/suppliers.
   - Sub-contractors/suppliers should be notified of the amount included in the valuation certified for the main contractor.

(iv) Value Added Tax (VAT)
   - VAT should be included in the value of works certified for the contractor; withholding tax should be deducted as appropriate by the client-paying office. Withholding tax should not be added or deducted from the valuation due.

(v) Negative valuation and amortisation of advance payments.
   - There may be a situation where, after deducting agreed parts of advance payment, the value of work executed, including on-site materials, is below the amount previously certified. With such an occurrence, instead of issuing a negative valuation, a valuation breakdown of work completed should be entailed and circulated for record purposes, and a "NIL" valuation should be indicated against the amount to be recommended for certification, thus indicating that the contractor has not done enough work to attract a payment recommendation.

     To avoid the occurrence of NIL valuations due to advance payment deductions, appropriate professional advice should be offered to the client regarding the amortisation of advance payments.

Usually, it is advisable to recommend a deduction proportional to the work completed by the contractor, provided such a deduction starts after the commencement of work. Amortisation of advance payments does not necessarily have to be in equal instalments.

Typical Valuation Form

PACE AND PARTNERS

(Registered Quantity Surveyors and Project Managers)

30 Asemota Street,
Off Igun Road,
Benin City, Phone: 052–275011
Edo State. 08023014101

PROJECT/WORKS———————————————————
CLIENTS ————————————————————————
CONTRACTOR—————————————————————
COMMENCEMENT DATE———COMPLETION DATE—————
CONTRACT SUM———ROUSED CONTRACT SUM—————
VALUATION NO. —————DATE OF VALUATION————
RETENTION —————%, LIMIT _________%, OF CONTRACT SUM
DATE OF ISSUE ——————————————————

1	16. Preliminaries	
2	Main contract	N
3	Nominated subcontracts	K
4	Nominated suppliers	
5	Materials on/off-site	
6	Add advance payment	
7	Value of completed work and materials on site	
8	Less: total amount of retention	
9	Add: fluctuation	
10	Add: release of retention	
11	Add VAT XE 5%	
12	Less: recovering of advance payment	
13	Gross amount payable to date (including VAT XE)	
14	Less: amount previously certified	
15	AMOUNT RECOMMENDED FOR PAYMENT	
	Amount in words:	

Signed and stamped for Pace and Partner
Typical Valuation Form

Example 1

The following examples demonstrate many of the details discussed previously in this chapter. They relate to a contract in which particulars are given as follows:

Location of site:	Ikoyi, Lagos
Description of works:	Four lock-ups at base level, with 8 flats over, and a basement car park and store.
Contract sum:	N190,211,481.24 k
Employer:	AP PLC, Ijora, Lagos

(Continued)

(Continued)

Main contractor:	Fhomo Nig. Ltd. Ikeja, Lagos
Architect:	AP properties
Quantity surveyor:	AP properties
Basis of contract:	Federal Ministry of Works, lump sum contract Quantities, 2005 edition
Date of possession:	September 6, 2018
Completion date:	June 29, 2020
Contract period:	20 months
Period of interim certificate:	One month

Abstract of Information from Contract Bills

Preliminaries	10,475,000:00
Substructure	9,487,847:55
Frames:	5,070,613:00
Upper floor:	9,303,606:50
Roof:	7.114.224:00
Internal and external walls:	5,260,000:00
Fittings and fixtures:	7,600,000:00
Wall finishes:	8,981,121:40
Floor finishes:	4,747,305:83
Ceilings finishes:	1,843,965:50
External works:	2,537,483:83
	72,421,167:61
Pc and provisional sums:	93,732,624:05
	166,153,791:66
Add contingencies:	15,000,000:00
	N181,153,791:66
Add VAT, 5%:	9,057,689:58
Estimated total cost:	N190,211,481:24

Record of the site visit made on October 20, 2018 is as follows:

Substructure 100% completed
25% of the work had been done in frames.

Materials on site:

Cement 400 bags @ N1,500
Sharp sand 15 trips @ N6,000
Granite 10 trips @ N25,000
225-mm-thick blocks 10,000 @ N100

Paid for utility board water connection N35,000
Paid for 3-phase metres @ N45,000

25% of the contract sum is to be included for payment in the first valuation and amortised five times.

Prepare interim valuation No. 1 (make all necessary assumptions where necessary).

Valuation No. 1 for AP PLC— October 20, 2018

	N K	N K
Main contractor's work:		
Properly executed to date:		
Substructure:	9,487,847:55	
Frames (25% of N5,070,613=):	1,267,653:25	10,755,500:80
Materials on site:		
Cement 400 bags @ N1,500 =	600,000	
Sharp sand 15 trips @ N6,000 =	90,000	
Granite 10 trips @ N25,000 =	250,000	
225-mm-thick blocks 10,000 @ N100 =	1,000,000	1,940,000 =
Statutory bodies:		
Utility board:	35,000	80,000 =
PHCN	45,000	12,775,500.80
Preliminaries:	190,211,481.24	
Contract sum:		
Less		
VAT 9,057,689.58		
Contingencies 15,000,000:00		
Prelims 10,475,000:00		
Pc and P Sums 93,732,624.05	128,265,313.63	
Value of measured work =	61,946,167.61	
% of work done and materials		
On-site to date:-		
Work done = 10,755,500.80		
Materials on site = 1,940,000:00 N12,695,500:80		
12,695,500 x 100		
61,946,167.61 1		
= 20.49%		
Value of prelims = 20.49% of N10,475,000 =		2,146,327.50
Gross valuation		14,921,828.30

(Continued)

(Continued)

Add: advance payment;		
25% of N190,211,481.24k		47,552,870.30 N62,474,698.60
Gross valuation, including advance payment		
Retention: 62,474,698:60k		
Less amount subject		
To nil retention;		
Statutory fee 80,000.00		
62,394,698.60		
Value of retention		
5% of N62,394,698.60		3,119,739.93
Net valuation including adv. Payment		N59,354,963.67
Add: VAT XE, 5%		2,967,748.18
		N62,322,711.85
Less: recovering adv. payment	N47,552,870.30 5	= N9,510,574.06
Gross amount payable to date, including VAT		N52,812,137.79
Less: amount previously certified		—
Amount recommended for payment		**N52,812,127.79**

Example 2

Prepare the relevant details of interim valuation No. 1, showing the amount due for payment for a project whose particulars are given below:

Contract:	A bungalow block of 4-bedroom flat for Messrs. John, A.Y.
Employer:	Auchi Technical College
Contractor:	Based on monthly valuations
Contract sum:	4,294,983:79
Limit of retention:	5%
Completion period:	6 months
Summary of the bills of quantities:	
A. Preliminaries:	192,750:00
B. Measured work:	2,797,710:75
C. PC and provisional sums/provisional items:	
(i) Electrical installation:	320,000:00

(Continued)

(Continued)

(ii) Plumbing installation:	290,000:00
(iii) External works:	240,000:00
Contingencies:	250,000:00
Value-added tax:	204,523:04
Contract sum:	4,294,983.79k

Notes:

1. 25% of the contract sum is paid as a mobilisation fee to be recovered in the first three (3) instalments of the payment.
2. The breakdown of the measured work shows that substructure = N590,000 is at 100% completion and frames and wall = N129,500 are at 30% completion, respectively, at the time of this interim valuation No. 1.
3. N30,000 was paid to the utility board for the connection.
4. Materials on site:
   (i) 35 bags of cement @ N2,300/bag
   (ii) 5 trips of sharp sand @ N9,000/trip
   (iii) 2 trips of granite @ N52,000/trip
   (iv) 1,200 nos of 150-mm hollow block @ N120/block

Answer No. 2

Valuation No. 1
For Auchi Technical College as at October 21, 2010

	N K
1. Preliminaries:	
% of work done and materials on site to date:	
Work done = 628,850 =	
Materials on site=373,500 =	
N1,002,350 =	
Value of prelims = 35.83% of N192,750:00	69,062.33
2. Main contract	
Main contractor's work executed to date:	
(i) Sub. N590,000 @ 100% completed =590,000	
(ii) F and W N129,500 @ 30% completed=38,850	628,850=
3. Nominated subcontracts	-
4. Nominated suppliers	-
5. Materials on site	
(i) Cement, 35 bags @ N2300 80,500 =	
(ii) Sharp sand, 5 trips @ N9000 45,000=	
(iii) Granite, 2 trips @ N52,000 104,000=	

(Continued)

(Continued)

(iv) 150mm block, 1200 @ N120 144,000= 373,500=	
6. Value of Completed Works and Materials on Site	1,071,412.33
7. LESS: Total amount of retention	53,570:62
	1,017,844.71
8. ADD: Fluctuation	
(i) Statutory connection fees	30,000:00
	1,047,841:71
9. ADD: release of retention	
10. ADD: VAT 5%	52,392:09
	1,100,233:80
11. ADD: Advance payment 25% of N4,294,983.79	1,073,745.95
	2,173,979.75
12. LESS: recovery of advance payment =	357,915.32
13. GROSS AMOUNT PAYABLE TO DATE	
(INCLUDING VAT)	1,816,064:43
14. LESS: Amount previously certified	-
15. Amt. Recommended for Payment	N1,816,064.43

Amount in words: One million, eight hundred and sixteen thousand, sixty-four naira, and forty-three kobo only.

Signed and Stamped for Pace and Partners.

5.2 FINAL ACCOUNTS

A final account is a statement of the final valuation under a contract. It is a control report comprising the actual cost (constructional sum) and the target cost (contract sum). The final account shows the changes experienced in a construction project and the impact made by such changes (Ameyaw *et al.*, 2015). The final account is prepared to enable the contractor to receive his final outstanding payment regarding the contract terms. This is pertinent because several adjustments are made to the contract sum during the execution of the construction project. The Federal Ministry of Works (FMW) Conditions of Contract Clauses 30(5) and 30(6) detailed the obligations of the Architect, contractor, and Quantity Surveyor regarding final account preparation.

(i) Before or within a realistic time after the practical accomplishment of the work, the construction contractor shall forward to the Architect supervising official (or, when trained, to the Quantity Surveyor) every single document required for preparing the final account.

(ii) The QS (subject to the contractor's compliance with the commitment in one (i) above) will set up the final account within the time of final measurement and valuation.

(iii) The Architect shall send a duplicate to the main contractor and significant extracts to the nominated sub-contractor when the final account has been completed.

5.2.1 Preparation of Final Accounts

The issues that ought to be managed in the final account to modify the contract sum in accordance are enumerated in the conditions; see FMW 30(5) and 30(6). The final account is normally arranged with the first contract sum as the beginning stage. The statement goes further to show what amount is deducted and what is included. The adjusted total sum will be shown at the end. This sum is the total amount that the client shall reimburse the contractor for the work executed. Note that the amount in the final certificate is the discrepancy between the total of the final account and the amount formerly stated as entitled under the interim certificate. The following are the components of the modification of the contract sum in the final account:

(i) Adjustment of prime cost sums
(ii) Adjustment of provisional sums
(iii) Variation account
(iv) Adjustment of provisional items or quantities
(v) Claims
(vi) Fluctuations

Sums to Be deducted

(i) Prime cost sums
(ii) Provisional sums
(iii) The value of work defined as provisional
(iv) Variations that are non-inclusion
(v) The amount allowable to the client under fluctuation Clauses 30 or 31 (whichever is applicable)
(vi) Any other amount that is required by the contract or conditions is to be subtracted from the total contract sum

Sums to Be Added

(i) Total amount of sub-nominated sub-contracts
(ii) Any amount due to nominated suppliers
(iii) The contractor's profit, or the amount referred to in (i) and (ii)
(iv) Any sum to be paid by the client relating to statutory fees, setting out the works, opening and testing, etc.
(v) Variations that are additions
(vi) Value of works executed against provisional sums or provisional quantities included in the bills
(vii) Any amount payable by the employer by way of reimbursement for claims
(viii) Any amount spent by the contractors because of loss or injury by fire or other perils where the risk is insured by the client and the contractor is eligible for a refund
(ix) Any sum to be paid to the contractor under the fluctuation clauses
(x) Any other sum obliged by the agreement or conditions to be included

In the preparation of the final account, a summary page is prepared. This page consists of a two-fold billing paper showing the contract amount. The individual total and deduction amount moved from every one of the succeeding segments to the final total. The contractor and Quantity Surveyor sign the base of the page. This is because the Quantity Surveyor is advised to give the contractor's Quantity Surveyor or contractor's agents a chance to be available when measurements are recorded on-site. This enhances transparency and all-inclusiveness. But delays in completing the final account signify additional expenses for the contractor. Thus, the client is worried about knowing his final financial commitment. A draft final account is a valuable system that maintains cost control for the agreement. It commences as building work begins and is updated at monthly or milestone intervals. From the draft account, a report on known and foreseen expenditures can be prepared and submitted to the Architect. This can be done monthly as a financial forecast of the probable final cost. This procedure empowers the account to be finalized after completing the building work.

5.2.2 Presentation of Final Accounts

The final account's presentation pattern is shown in appendices 1–4. The contents of which are described hereunder.

5.2.2.1 Appendix 1—Summary of Account and Final Statement of Account

This section of the account is in two parts: the summary of an account, which shows the final contract sum, and the final statement of account, which shows the first balance because of the contractor on the final certificate. The balance due is arrived at by deducting the total amount previously paid to the contractor from collecting the total amounts due on all aspects of the work. The final contract sum, normally indicated as the total on the account summary, is agreed upon after adjusting prime cost and provisional sum items, re-measurement of provisional quantities, measurement of variations, valuation of fluctuations, and ascertainment of contract claims. The "final statement of account" states that the "final balance" outstanding to the contractor is the final payment due on the contract.

5.2.2.2 Appendix 2—Adjustment of PC and Provisional Sums

The account for prime cost and provisional sums contained in the contract are presented for adjustment on the account summary using the "Omission" and "Addition" tabulation methods. The total of the omissions and additions is carried forward to the account summary (Appendix 1). The adjustment is made to the account's summary by deducting the omissions and adding the additions.

5.2.2.3 Appendix 3—Summary of Variation

Accounts for works carried out under variation orders and instructions to add new works or substitute existing works with new ones are included in this section. Works executed under day-work instructions may be included in this account section. This account is carried forward to the summary of accounts as a net addition or omission.

5.2.2.4 Appendix 4—Fluctuations and Claims

Accounts for labour and material fluctuations and other contract claims are presented in this section. In general terms, this section is often erroneously termed an increased cost account due to a tendency for the account to comprise mainly "Addition" items, but it should be noted that fluctuation could also lead to a decrease in costs. Liquidation and ascertained damages are omission items under contract claims, while claims for loss and expense are additional items.

Example 1 of Final Account Source: NIQS Practice Note 5

APPENDIX 1

A. Summary Of Accounts

	N
To value contract dated January 24, 1998	800,500,345.00
LESS: PC and provisional sums (omission, appendix 2)	12,875,345.00
	787,625,345.00
ADD: works covered by P.C and provisional	
sums (Additions, Appendix 2)	11,528,321.00
	799,153,666.00
ADD: Value of variations included in variation orders (Nrs. 1–38) and	4,249,334.00
remeasured works (Appendix 3)	803,403,000.00
ADD: Value of fluctuation and claims (Appendix 4)	25,052,555.00
.Final contract sum	**828,455,555.00**

B. Final Statement Of Accounts

I. Final Account Sum	
(As on summary of accounts)	828,455,555.00
II. Less: Total previous payment	
(As on schedule of payment)	809,489,655.00
Final statement of accounts	
Final balance outstanding and due to	
Messrs. Construct well Ltd, on Final Certificate	**18,965,900.00**

APPENDIX 2

Summary of Adjustment of Prime Cost And Provisional Sums

	Omissions	Additions
1 Contingency sum(items KN3/2A)	5,000,000.00	
2 Structural steel work (itemsKN3/P)	2,500,000.00	4,434,100.00
3 Plumbing installation (itemsKN3/4A-G)	1,102,000.00	2,485,230.00
4 Electrical installation (items KN3/9C-K)	3,360,000.00	4,001,200.00
5. Purpose made security installation XE (KN3/10M)	1,000,000.00	607,791.00

Summary adjustment of		
PC AND PROVISIONAL SUMS		
Carried to account summary	**12,875,000.00**	**11,528,321.00**

APPENDIX 3

Summary of variations covered by Variation order Nos. 1–38

Variations	**Omissions**	**Additions**
(1) Replacement of wooden frames with steel frames	398,500.00	755,320.00
–	–	–
–	–	–
–	–	–
(34) Modification of storm water drainage	–	320,405.00
–	–	–
–	–	–
(36) Use emulsion paint in lieu of gloss on lavatory walls and ceilings	121,300.00	84,400.00
Etc. up to V.O. Nr. (38)	–	–
Remeasured works	–	–
Remeasured substructures	3,827,411.00	3,974,822.00
–	–	–
Etc.		
Total omissions/additions	5,013,433.00	9,262,777.00
		5,013,443.00
Value of variations		
Carried to account summary		**4,249,334.00**

APPENDIX 4

Fluctuations and Contract Claims

	Omission	**Additions**
Labour increases		
Up to September 1998		
Tradesmen 600 hours . . . per month 36,000)		
Labourers 2300 hours . . . per month 54,440)		90,440.00
September 1998 to April 1999		
Trade men . . .		
Labourers . . .		
May 1999 to July 2000		134,345.60
Trade men . . .		
Labourers . . .		
–		1,803,453.98

(Continued)

(Continued)

–	–
etc.	
Material price increases	
Actual price basic price	
Material A 31,343,200.00 28,245,675.00	3,097,525.00
Material B 15,443,100.00 13,403,080.00	2,040,020.00
–	–
–	–
etc.	–
Claims	
–	–
–	–
Etc.	
Fluctuations and claims	
Carried to accounts summary	25,052,555.00

Example 2 of Final Account

Prepare the relevant details of the final account showing the amount due for payment for a project whose particulars are given below:

Contract:	Proposed office complex for XYZ and Company Limited with an underground car park
Employer:	XYZ Properties Limited
Contractor:	Project Delivery Associates;
Contract sum:	N105,000,000.00;
Limit of retention:	5%
Completion period:	52 weeks
Summary of Bill of Quantities	
A. Preliminaries	8,100,000=
B. Measured work	51,426,400=
C. Nominated sub-contractor	
i. Aluminium components	5,500,000=
Add for profit 5%	275,000=
Attendance	275,000=
ii. Plumbing installation	4,000,000=
Add for profit 5%	205,000=
Attendance	82,000=

(Continued)

(Continued)

iii. Electrical installations	6,100,000=
Add for profit 5%	305,000=
Attendance	122,000=
iv. Lift installations	4,500,000=
Add for profit 5%	225,000=
Attendance	225,000=
v. Suspended ceilings	950,000=
Add for profit 5%	47,000=
Attendance	23,350=
D. Nominated Suppliers	
i. Iron monger	2,800,000=
Add for profits 5%	140,000=
ii. Sanitary fittings	1,000,000=
Add for profit 5%	50,000=
E. External works	8,548,750=
Contingencies	5,000,000=
Value added tax	5,000,000=
Contract sum	N105,000,000

Notes:

1) The contractor tendered for aluminium components and submitted a quotation of N5,125,000, which was the lowest and subsequently accepted.
2) The following quotations were accepted in respect of items included in the Bill of Quantities: nominated sub-contractors
   a) Aluminium components 5,125,000=
   b) Plumbing installations 4,200,000=
   c) Electrical installations 5,700,000=
   d) Lift installations 4,300,000=
   e) Suspended ceiling 1,010,000=

   Nominated supplier
   a) Iron monger 2,650,000=
   b) Sanitary fittings 900,000=
3) A claim of an extra value of N315,000= has been made regarding the increase in the price of iron monger.
4) Variations and additional works have been valued at N3,515,000= as an addition to the contract sum.
5) The external work has been re-measured and valued at N9,125,000=
6) Previous payment, including the value added tax, has been made to the value of N90,252,000=

Summary of Final Accounts

Dated May 20, 2020.	Omission	Addition
	N K	NK
Contract sum as contract dated		105,000,000=
Contingencies	5,000,000=	-

(Continued)

(Continued)

Adjustment of prime cost and provisional sum (appendix 1)	26,924,350=	23,885,000=
Variation account		3,515,000=
Fluctuation and claims		315,000=
Adjustment of provisional sum	8,548,750=	9,125,000=
	40,473,100=	141,840,000=
LESS: Omission		40,473,100=
		101,366,900=
LESS: payment on account		90,252,000=
Amount of final certificate		N11,114,900=

Signed
For and on behalf of Project Delivery
. . . .

Signed
for and on Behalf of consultant Q.S.
. . . .

APPENDIX 1
Summary of Adjustment of Prime Cost and Provisional Sums

		Omissions	Additions
1.	Aluminium components	6,050,000	5,125,000
2.	Plumbing installations	4,387,000	4,200,000
3.	Electrical installations	6,527,000	5,700,000
4.	Lift installations	4,950,000	4,300,000
5.	Suspended ceiling	1,020,350	1,010,000
6.	Iron mongery	2,940,000	2,650,000
7.	Sanitary fittings	1,050,000	900,000
	Total	**26,924,350**	**23,885,000**

Note:

- No appendix is required for fluctuation.
- No appendix is required for variation accounts.
- No appendix is required for daywork; no day work at all.

Example of Final Account (Source: Bowen Partnership)

Project: Ebek's Water Factory
Employer: Engr Augustine and Lady Elizabeth Ebekozien
Contractor: Meek Construction Ltd.

Final Account (Draft)

		N	N
Original contract sum			64,905,489:00
Less			
1.1	Provisional sums in the contract (Appendix A)	2,710,000:00	
1.2	Provisional sum expunged from the contract (Appendix A.1)	3,123,875:00	
1.3	Provisional quantities in the contract (Appendix B)	19,531,475:81	
1.4	Prime cost sums in the contract (Appendix C)	10,735,000:00	
1.5	Profit and attendance on P.C. items in the contract (Appendix D)	738,412:34	
1.6	Items not executed in the contract (Appendix E)	34,484:46	36,873,247:61
			28,032,241:39

Add			
1.1	Actual expenditure against provisional sums (Appendix F)	725,000:00	
1.2	Attendance on marmoran painting (Appendix F.1)	234,290:63	
1.3	Actual expenditure against P.C. Sums (Appendix G)	11,857,179:30	
1.4	Profit and Attendance on actual cost of P.C. Items (Appendix H)	789,872:12	
1.5	Actual expenditure against provisional quantities (Appendix J)	23,003,180:90	
2.6	Approved Variations (Appendix K)	2,643,130:27	
2.7	Additional cost of Prelims due to		
	Extension of time (Appendix L)	566,333:95	
2.8	Fluctuation (net labour and net material increase—Appendix M)	10,797,544:56	50,616,531:73
			78,648,773:12

Add		
3.1	Value added tax (VAT) – 5%	3,932,438:66
	Total Final Account Figure	82,581,211:78

5.3 INTERIM FINANCIAL STATEMENT

This is a summary valuation of the total cost of the construction works, mirroring the financial implications of all instructions and modifications determinable on projects up to the time of the statement. Keep a precise account of the contract price, from the initial cost to the final price. Also, keep a record of factors affecting the project agreement financially as they unfold, even before issuing instructions. This will convey to the client the financial commitment up to the end of the construction project.

5.3.1 Summary of Financial Reports

This is the financial statement form, which is a summary that contains the following:

A) Initial contract sum
B) Client-revised requirements
C) Fluctuations
D) Variation (other revised requirements) and adjustments
E) Estimated (anticipated) final cost
F) Net: Cost overrun/anticipated savings

Details of each item on the financial statement summary shall be included in the backup sheets, giving a necessary breakdown as the presenter deems appropriate but following the corresponding format. For example, the details could be presented in broad elemental form. Items that appear in the backup sheets for which breakdowns may be provided are:

i. Additional or revised requirements by the client
ii. Fluctuation
iii. Adjustment of provisional sums
iv. Adjustment of prime cost sums
v. Adjustments remeasured provisional quantities
vi. Effects on projects: the cost of instructions to vary or substitute works and/or specifications
vii. Anticipation variation in cost due to impending instruction not yet formalised
viii. Direct loss and expense

5.3.2 Other Information for Financial Reports

i. Liquidated damages
Where the employer decides to claim liquidated and ascertained damages, the amount shall be indicated, and the completion duration assumed as the basis for determining the number of damages due should be stated. Damages shall not be computed for the period covered by the extension of time.
ii. Claims submitted by a contractor but not included in the financial statement summary shall be separately stated for awareness purposes and necessary attention.

5.4 DAYWORK

Daywork reimburses the construction contractor for the prime cost of all materials, labour, and plant utilised in executing the work, with a percentage extra to cover for profit and overhead. FMW Contract 11.4.1 states that daywork is restricted to extra or alternated work of variation that cannot be properly measured and valued. This is because it encourages wastage on the part of the contractor. This results in greater expenses for the client than when a measure and value basis are employed. Clause 11.4.1 of the FMW form stipulates that the prime cost of work (computed following the definition of prime cost of daywork executed under a building contract . . .) together with a . . .%, adding to the prime cost of labour, material, and plant at the rate inserted by the contractor in the agreement bills. This explains the extent of what is inserted in the prime cost in every one of the three divisions of labour, material, and plant, other than direct cost. Daywork can be set out in the tender document in any of the two ways:

(A) The Form of Tender on the contract bill can include the column for the schedule of daywork where space will be provided for the construction contractor to insert a percentage against the prime cost of labour, material, and plant (a typical column usually denoted appendix "in the bill as shown below).

"Appendix A"

Schedule of Daywork Rate: Following Clause 11(4)(c)(i) of the FMW Condition of Contract, the contractor shall be paid the daywork rate, any extra or alternated rate which cannot be valued by measurement. The contractor shall insert the percentage that will then be required to add up to the basic rate and cost of labour, materials, and plant as presented below:

i. Labour
   To the basic net rate of labour at the hour of any such daywork, an extra of . . .% is incorporated for task work and incentive plan, tools, scaffolding, supervision, insurance, holidays with pay, transport, profit, and overhead.
ii. Materials
   To the basic net invoiced cost of material conveyed to the site at the hour of any such daywork, an extra of . . .% is incorporated for emptying and storage as required, profit, and overhead.
iii. Plant
   To the basic invoice rental price of plant, machinery, and equipment employed on such daywork, an extra of . . .% which is to include insurance, maintenance, fuelling and consumable stores, transport to and from the site, profit, and overhead.

FMW Clause 11.4 (c)(i)(ii) stated that daywork is priced at the rate inserted in any schedule of daywork. This is modified as agreed to precede the work between the client and a representative committee of building trade employers for the area affected. Where there is no such concurred schedule, the contractor will be permitted a net cost plus 25%. Daywork carried out after practical completion (during the defect liability period) could justify different percentages as plant, labour, etc. Rates and percentages set out in a daywork schedule in the above form do not affect the tender. Therefore, the tenderer is tempted to make such a rate high (a disadvantage of the schedule form). Such a rate may be negotiated for a reduction during the occurrence of the work. From experience, getting the contractor to agree to such a reduction is difficult.

(D) Daywork could be priced in the contract bill and reflected as a sum in the completion. The bill may be drawn up as follows:

Bill No . . . Daywork

	N K	N K
Provisional		
Daywork could only be permitted when work cannot be properly measured and valued, according to Clause 11.4 of the FMW and Clause 13.5.4 of the JCT form. The contractor shall add the percentage addition on prime cost he requires on the following and include the total so derived in the tender.		
Builder's work		
Labour		
Provide the sum of N100,000 (one hundred thousand naira only)		
<u>Add</u>: For percentage addition ....%		
<u>Material</u>		
Provide the sum of N150,000 (one hundred and fifty thousand naira only)	100,000=	
<u>Add</u>: For percentage addition ....%		
<u>Plant</u>		
Provide the sum of N60,000 (sixty thousand naira only)	150,000=	
Add: For percentage addition ....%		
<u>Specialist trade work (specialist stated)</u>		
<u>Labour</u>		
Provide the sum of N40,000 (forty thousand naira only)	60,000=	
<u>Add</u>: For percentage addition................%		
<u>Material</u>		
Provide the sum of N30,000 (thirty thousand naira only)	40,000=	
Add: For percentage addition. . .%		
<u>Plant</u>		
Provide the sum of N20,000 (twenty thousand naira only)	30,000=	
<u>Add</u>: For percentage addition. . .%		
Carried to Summary	20,000=	

With this method, the tenderers will probably insert a not-too-exaggerated percentage, as they will be affected if tendered. Clause 11.4 (a)(ii) of the FMW contract provides that where there is no schedule of daywork charges, the contract shall allow the net cost of labour and material employed in daywork plus 25% of the cost to cover profit. This will include all contract expenses except the cost of labour and materials in the sub-clause. This percentage covers head office, site overhead, small tools, etc. Also, under this system, the plant and operative employed in daywork will be paid for at the rate concurred with the agreement preceding its execution. When work is a specialist, the definition of prime cost of daywork stated by the appropriate institution and professional body at the date of the tender should be adopted for settlement. The percentage addition of the prime cost shall be that already quoted by the contract bills.

5.4.1 Recording Daywork

The record of time spent daily in such daywork, the workmen's name (only where the Architect/supervising officer requires it; the FMW contract refers), and the materials and plant employed should be recorded in vouchers.

A typical example of daywork recorded on site.

Fhima Nigeria Limited
(Building, Civil and General contractors)
2 Toyin Street, Apapa, Lagos.

Project Title: Construction of Shopping Mall for AP at Ikoyi, Lagos.
Work commenced: 20–10–2019
Work Completed: 14–5–2020
Daywork Sheet. No ....Date: 20–05–2020
A1 No .... omit window marked 2 in room number five, opening attached as required to form door opening, provide and fix 838 x 1981 mm door frame and door, fix mortice lock and lever handle.

Labour	Hrs	Rate	N	Materials	Qty	Rate	N
Mason	16	2500	40,000	450×225×150 mm blocks	20	140	2,800
Carpenter	6	2500	15,000	Mortar	1/10 m^3	9,000	900
Labourer	10	1800	18,000	Precast lintel 1138 mm long	1nr	5,400	5,400
Painter	8	2500	20,000	838×1981 mm flush door	1nr	6,000	6,000
Plant				Door frame set	1nr	1,300	1,300
				Undercoat paint	1/2 litre	260	130
				Finishing coat paint	1/2 litre	260	130
				Pair steel hinges	2nr	180	360
				Pair bolt	2nr	180	360
				Fixing cramp	6nr	150	900

The information shown on this sheet has been verified.

------------------------ ------------------------

Site Agent Architect/Representative.

5.4.2 Daywork Account

AI No	Rate	N	N
Issued date . . .			
Expected date . . .			
Daywork sheet No . . .			

Description Omit window marked 2		
Labour		
30 hours craftsmen	2,500	75,000
10 hours labourer	1,800	18,000
		93,000
Percentage addition (say) 120%		111,600
		204,600

Material			
20 nrs block	140	2,800	
0.1 m³ mortar	9,000	900	
1138-mm long PC Lintel (1nr)	5,400	5,400	
Flush door (1nr)	6,000	6,000	
Door frame set (1nr)	1,300	1,300	
2 pairs hinges	180	360	
Pair of bolts	180	360	
Fixing cramps (6nos)	150	900	
½ litre undercoat paint	260	130	
½ litre finishing paint	260	130	
		18,280	
Percentage addition (say) 150%		27,420	45,700
Total addition carried to summary			**250,300**

Note: The omission of the window complete with frames and lintel, block wall, and painting would have been measured and valued at the bill rate. The total of such omissions will be carried to the summary or daywork "sheet" and delivered to the Architect for rectification not later than one week after the work is done. The signature of rectification by the clerk of work certified the correctness of the time and materials recorded, not the correctness of any rates included. The Quantity Surveyor will verify the correctness of the rate and determine whether items that could have been properly valued by measurement are included in the daywork voucher.

5.5 FLUCTUATIONS

Clause 31 of the FMW contract provides an adjustment to the contract sum for increases or decreases which may occur during the progress of the work in labour rates and material prices. This only applies to construction contracts whose completion period exceeds twelve (12) months. The contract sum shall be considered to have been calculated based on the following:

(a) LABOUR: The rates of incomes and other emoluments and expenses to be paid by the contractor in line with the rates and wages fixed by the NJIC and the Ministry of Labour and Productivity prevailing at the time of tender and
(b) MATERIAL: The current prices at the time of tender of materials and goods listed in the basic price paid on plant or overheads and profits.

5.5.1 Procedure

As money becomes due to the contract under the fluctuation clause, it should be included in interim certificates. Since only net amounts are allowed, they are usually not subject to reduction for retention purposes. The contractor should be instructed to submit claims at regular intervals. All timesheets should be endorsed by the clerk of works at the end of each week as confirmation that the people listed worked the times stated. The checking of fluctuations claims can be rather tedious sometimes. Therefore, it is better done regularly during the contract period, allowing timesheets and invoices to pile up for checking at the end of the job. Changes in wage rates usually occur nationally and are announced in the press in advance. There should be no dispute as to when the fluctuation took place. The procedure is not straightforward in connection with materials prices and may differ from one supplier to another.

Only materials entered in the basic price lists will rank for adjustment under this clause. Once the contract has been signed, the basic price list cannot be altered except by mutual agreement. One of the responsibilities of the Quantity Surveyor during the scrutiny of a tender under consideration is to check the basic price list and satisfy himself that the rates are reasonable. Any mistake made here could involve the employer at considerable expense later. The tenderer should substantiate his basic prices with copies of quotations from potential suppliers. The Quantity Surveyor should acquire photocopies, if possible, of these quotations so that if a difference is subsequently used, the Quantity Surveyor will ascertain whether the price increase is not due merely to a change in the supplier. The basic price should preferably be for material delivered to the site. The contractor should notify the Architect within a reasonable time when fluctuations occur. Any "ex-works" prices should be carefully noted.

When the material fluctuation claim is complete, the Quantity Surveyor must check the quantities of the various materials against those in the bill of quantities. Also, he should check the variation account to ensure there is no unreasonable discrepancy. For example, if the price of sandcrete blocks has varied and the contractor's claim shows that all 20,000 blocks were delivered to the site, the total area for blockwork contained in the bill and any variation orders should be found, and the number of blocks required for such an area calculated and compared with the claim for 15 nos excess of the required number. The contractor must explain how the discrepancy has arisen. To carry out such a check for a material that fluctuates in price, say halfway through the contract period, it will be necessary to extract quantities from all the invoices, including those issued before the fluctuation occurred. The main reason for this check is to make sure materials used on one contract are not changed against another. This implies that the employer is paying for materials used on his project.

As already noted, only materials entered in the basic price list are to be included in the fluctuation claim. The basic price of the material listed must not be set against the purchase price of another similar material. For example, suppose the Architect orders a change in the type of cement to be used, from ordinary to rapid hardening, and only the former appears in the basic price list. In that case, the change must be treated as a variation, and a new rate must be agreed upon for concrete, including the rapid hardening of cement. When checking the basic price list, attention should be paid to any minimum quantities stated alongside certain prices. Frequently, materials bought in bulk, that is, in large quantities, are sold at a lower unit price than in small quantities. A contractor sometimes uses materials from his stock that he purchased before the contract commenced. If these materials increase in price during the contract period, the contractor will not be paid the current market price. Such materials are treated like any others bought before an increase takes place. Conversely, if a contractor uses materials from his employer, he cannot recover the amount of such a decrease.

Additional Considerations

(i) Categories of labour

The fluctuation in wage rates applies to the following categories of workers.

(a) Operatives are employed directly on site.

(b) People engaged elsewhere by the construction contractor in producing materials and merchandise for the works, for example, joinery shop workers.

(c) People, for example, timekeepers and site surveyors, who, connected by the contractor on-site, are not workpeople. The sum regarding this group of individuals is equivalent to that of a craftsman compared to the time spent on construction sites per week.

(d) Workers engaged by domestic (non-nominated) sub-contractor rank for the recuperation of fluctuations.

(ii) Exclusions

The following works are excluded from the fluctuation provision:

(a) Work paid on a daywork basis.

(b) Work done by a nominated sub-contractor and materials and merchandise provided by a nominated supplier. These amounts are taken care of in the clauses regarding these people.

(c) Work done by the construction contractor is the subject of prime costs, for which the contractor presented a tender for a reason connected to (b) above.

(d) Adjustments in the value-added tax were made by the construction contractor to the client.

(iii) Fluctuation after the expiry of the Contract Period

Fluctuation sums are recoverable until the date of authentic completion. These fluctuations are "frozen" at the employment levels towards the completion date (as expressed in the agreement).

(iv) Other non-recoverable fluctuations

The following should be noted:

(a) Where a construction contractor pays employed employees above the nation's negotiated rates, the difference cannot be recovered as fluctuations.

(b) The non-production hours of additional time (in any event, when explicitly approved) should not be included in allowable fluctuation payments.

(v) Formula—Method of fluctuation recovery

The discussion above concerns the conventional technique of fluctuation recovery, which is to establish the real sum of fluctuations in costs and prices within the extent of the suitable classes of the agreement. The amounts recorded by the traditional method are far less than the actual expense acquired by the contractor. In contrast to the conventional technique, the object of the NEDO formula method is to calculate a sum or sums that will remunerate the parties to the agreement for losses incurred due to increases and decreases in materials. However, fluctuations in price are calculated on a month-to-month basis and added to the month-to-month valuation. **Note that retention is not held on fluctuations, as the contractor may not add a profit to the fluctuations claim.** Most standard forms of contract contain a fluctuation clause. Even though FMW did not give details of how fluctuations should be calculated, JCT 2011, a type of standard form of the method to be utilised by reference to schedule 7, options A, B, and C, where neither option B (labour, materials cost, and tax fluctuations) nor option C (formula adjustment) is chosen and expressed in the agreement particulars, option A (contribution levy and fluctuations) applies.

Option A—Firm price contracts (contributions, levy, and tax fluctuations)

One of the attributes of the fluctuation clause is the engagement of a base date from which any fluctuations may be calculated. The base date is decided between the parties to the agreement and is binding. This preference permits the construction contractor to be reimbursed for increases in the following items:

(i) Contributions, levies, and taxes payable regarding workpeople on site.

(ii) Contributions, levies, and taxes payable regarding workpeople off-site, producing materials or goods for the site.

(iii) Duties and taxes on fuels, electricity, and material, etc. When this is applied, only increases in taxes and levies, for example, VAT, are reimbursed. These were identified in JCT 2011. Also, Clause A12 of Schedule 7 offers for a rate to be embedded in the agreement particulars by the construction contractor. This will be applied to all fluctuations and allow for cost inflation relating to some preliminary items, overheads, and profit.

Option B—The traditional method (labour and materials cost and tax fluctuations)

This option allocation is more risk to the employer as he will be liable for paying increases that occur in the following situations: as option A, plus:

(i) Increases in rates of wages for workers on- and off-site and site workers following applicable wage-fixing bodies.
(ii) Transport charges.
(iii) Materials, electricity, and fuels.

Option C—Formula Adjustment

The general technique of price alteration in building contracts uses formulas to calculate the adjustment. In contrast to the conventional technique of repayment of fluctuations, the object of the formulae technique is to calculate sums, which will compensate the parties to the agreement for losses incurred due to an increase or decrease in costs. The formulae technique is unified in the "with quantities," "without quantities," and "with approximate quantities" variants of the JCT form, employing fluctuations (option C). The JCT form refers to fluctuations (option C.1.1.1) in a separate document called the formulae (rules). These rules lay out the formulae and define how they are used. For the formula, construction work is divided into 42 workgroups. The formula for adjustment of the value of work allocated to the workgroup under Part 1 of these rules of the valuation period up to and including the valuation period in which the date of practical completion occurs is as follows:

$$C = \frac{V\ (Iv - Io)}{Io}$$

where

C = the amount of adjustment for the construction work category to be paid to or recovered from the construction contractor.
V = the value of work in the work category for the valuation period.
Iv = the index number for the work category for the month during which the midpoint of the valuation period occurred.
Io = the index number for the work category for the base month.

These are separate formulae for specialist engineering works such as:

(i) Catering equipment installations for which specialist formula index numbers are published in the monthly bulletin.
(ii) Structural steelwork installations
(iii) Electrical and lift installations
(iv) Heating, ventilating, and air conditioning, and sprinkler installations

The following items are excluded from the formula adjustment:

(i) The amount for work is valued as daywork under the conditions.

(ii) Amounts for articles manufactured outside the country in which the employer requires the contractor in the contract documents to purchase and import or have imported for direct inclusion into the projects.
(iii) Amounts exempted from the formula amendment by the operation of rules 14 and 15.

The FMW and JCT suite of contracts makes the most provision for the recuperation of fluctuations in all the standard forms of contracts in use. Also, the government construction/works (GC/works/I) have no fluctuation provision at the other end of the spectrum. The assumption is that the agreement will have a fixed price. The Federation International Des Ingenieurs-Conseils (FIDIC) and the Institution of Civil Engineers (ICE) 7th edition are other contract forms.

Example of Calculation of Fluctuations (Traditional Method)
Labour and Material Price Fluctuations

	Omissions		Additions
	N		N
Labour increases			
Up to April 2010			
Tradesmen 400 hours@ N . . . month 24,000			
Labourer 800 hours@ N . . . month 14,000			38,000=
June 2009 to March 2010			
Tradesmen			
Labourers . . .			25,000=
March 2008 to May 2009			
Tradesmen . . .			
Labourers . . .			40,000=
Etc.			
Material price increase			
	Actual price	basic price	
Cement	3,101,200	2,700,100	401,100=
Reinforcement	7,170,000	4,500,190	2,669,810=
Granite	1,750,000	879,000	871,000=
–	–		–
–	–		–
–	–		–
Etc.			
Fluctuation carried to accounts summary			N4,044,910=

5.6 PRIME COST AND PROVISIONAL SUMS

5.6.1 Prime Cost Sums

The term prime cost can be defined as "the total cost to the contractor of buying materials, goods, and components, using or hiring plants, and employing labour." This is represented in a lump sum in the contract bill and will be executed by a nominated sub-contractor or supplier. The general procedure is to deduct prime cost sums and substitute the corresponding exact sums expended at the end of the construction project. The following details should be considered before the matter allows for adjustment:

(a) Discount

Discounts are sums by which offered or expressed costs or prices may be lessened when making a payment. This depends on the fulfilment of specified conditions. This may be a cash discount or a trade discount.

(i) Trade discount: This discount is conventionally permitted for the prices of the standard price list of materials and goods. This is often done to circumvent many revisions of such a list. Also, some suppliers will permit a larger discount for regular customers or those who buy large quantities.

(ii) Cash discount: This is allowed off-invoice prices if the contractor pays within a defined time. For the sub-contractor, this period is seven days after receipt of payment by the main contractor on an Architect's certificate. This passes on to the sub-contractor. The amount he allowed against the prime cost sum must include the appropriate cash discount.

Under the standard form, all discounts, excluding cash discounts, must be subtracted from any amounts due to sub-contractor and suppliers. Also, a cash discount of above 2½% for sub-contractors and 5% for suppliers must be subtracted. This means that the employer has the benefits of all discounts except the specific percentages for a cash discount.

(b) Net price and incorrect discount

At times, sub-contractors and suppliers quote net prices, that is, prices exclusive of a cash discount or prices inclusive of a cash discount, but at an incorrect percentage. The provision of the standard forms offers the contractor the obligation to pay a suitable discount. Also, offer the client the right not to pay more than the appropriate discount. Suppose the Architect instructs the contractor to place an order against a quotation that includes a correct discount or no discount. In practice, such stringent law enforcement may not be very common. It should be noted that the stipulated percentages for cash discount apply to the total cost of materials or commodities supplied, including any tax or duty and the charge of packaging, carriage, and delivery, and to the total value of sub-contract works, including any daywork and fluctuating amounts. Discounts should not be expected from statutory undertakings because they do not normally give cash discounts.

Correction of wrong discounts
Example of corrected discounts

A nominated sub-contractor tendered an invoice for N1,560:00 for wall tiling supplied and laid, the invoice being marked "net monthly account." The invoice ought to have included a 2½ cash discount, and to correct the error, 1/39th should be added because the percentages are always deducted from gross totals.

	N	K
Correction: Amount of invoice	1560:	00
Add: 1/39th	40:	00
	N1600:00	

(check: 2 ½ of N1600 = N40.00)

The sum of N1,600 will be paid by the client to the contractor, who will retain 2½% of the sum (i.e., N40.00) and pay the balance of N1,560 to the sub-contractor.

(c) General contractor's profit

Prime cost sums are regarded as selective of the general contractor's profit. The arrangement is made in the bills for an extra, which is calculated as a percentage of the prime cost sum. The tenderer is expected to state the percentage that he requires. When the omission of prime cost sums amends the contract sum, the replacement of real costs will be incurred. The profit sum will also be adjusted pro-rata. At times, the profit additions may be shown as lump sums without any indication of the calculation technique. Also, such sums should be adjusted pro-rata to the actual costs. If the client pays a subcontractor directly for any reason, the general contractor is still entitled to the profit addition. This is a management fee for arranging, organizing, and taking charge of the sub-contract for arranging and checking the supply of goods and materials, as the case may be.

(d) Attendance on nominated sub-contractor

A general contractor must "attend upon" a nominated sub-contractor, as he is accountable for their work. He integrates the sub-contract works into the main contract works. When there are bills of quantities, individual items are given for general attendance and other attendances (special attendance). General attendance is defined as "the utilisation of temporary contractor roads paving and paths, standing scaffolding, standing power-operated hoisting plant, the provision of temporary lighting and water supplies, cleaning away rubbish provision, provision of space for the sub-contractor own offices and storage of his plant and materials and the use of mess rooms, sanitary accommodation, and welfare facilities."

Other attendances include such facilities as scaffolding, temporary roads, and hard standing; unloading and positioning heavy and large items; storage accommodation; and the provision of power supplies. The prices included in tenders are usually treated as fixed sums. This is unlike the profit;

additions are not usually adjustable pro-data to the actual cost of the sub-contract work. This is because the kind and amount of work or services involved in the attendant remain the same. Only the client of sub-contract work can vary from what was originally envisaged to the extent of affecting the nature of the attendant, should the attendant amount vary. Sometimes, contractors show prizes for attendant items as a percentage of the prime cost sum in the same way as the profit items. In such an instance, the amount should be treated as a lump sum, and the Quantity Surveyor must agree on deleting the percentage rate with the contractor before signing the contract.

Note: A general contractor may execute work for which a prime cost sum has been inserted in the contract document or arisen under an AI. This is allowed by the standard forms, but with limitations as follows:

i. The work should align with the same approach the main contractor adopts if the project is approved. This is subject to the Architect's Architect concession.

ii The contractor should give notice preceding the drawings of the agreement that he wishes to complete such tasks. In the case of the emergence of a prime sum because of the AI, notice should be given as soon as the instruction is received.

iii The Architect must be ready to oblige a tender from the contractor, except where it is proven that the specialist sub-contractor will deliver regarding the task performance. The Architect can then consider the specialist sub-contractor to do it, in which case he has more confidence to provide a satisfactory standard of work.

Whether a contractor is eligible to be reimbursed, the profit addition for the execution of work by a prime cost sum depends on whether the sub-contractor tender was submitted in the competition. If awarded competition, the contractor is eligible to profit just as if a sub-contractor does the work. If not, the contractor must include the full amount and profit in his quotation for the work. However, attendance is exempted from the payment because a general contractor cannot attend upon himself. Other attendances, which involve extra expenses, may be allowed.

5.6.2 Provisional Sums

Provisional sums can be defined as "lump sums included in the contract sum to cover the cost of parts of the work which cannot be completely predicted, well-defined or detailed at the time tendering documents are presented." When the contract sum is adjusted, as usual, the provisional sums will be subtracted, and the total value of the work executed will be replaced. The construction contractor may be requested to present a tender for executing the work, and the amount tendered will be set against the right provisional sum in the accounts. The general contractor executes works covered by a provisional sum, and no item of profit or attendance is attached to such sums in the BOQ.

However, the Architect can sometimes transform the provisional sums into prime cost sums. This can be awarded to subcontractors or suppliers. But the reasons for such action must be compelling enough; otherwise, the contractor must object to such an arrangement. The contractor will be eligible for profit and attendance fees in this case. Where the actual value of work exceeds the provisional sum, the excess amount is treated as a net extra to the contract sum. Where the surplus is large, however, a contractor might claim an extension of time and reimbursement of any additional loss or expense.

Practical Example of Adjustment of Prime Cost and Provisional Sums General Account

Summary of Adjustment of Prime Cost and Provisional Sums

		Omission	Addition
1.	Contingency sum (Item P5/P21)	200,000	
2.	Structural seel work (Item M3/P14)	1,245,000	2,100,000
3.	Plumbing installation (Item L10/P10)	2,450,000	2,500,500
4.	Electrical installation (Item D7/P8)	3,400,000	5,100,000
5.	Purpose security Install (Item G7/P7)	1,500,000	2,500,000
		8,795,000	12,200,500
			8,795,000
Net addition: Carried to summary			3,405,500

Practical Example of Adjustment of Prime Cost Sums

	Omissions	Additions
Electrical installation		
P.C sum as item 145A	12,000,000	
Profit as item 145B 5%	600,000	
General attendance as item 145C	300,000	
Other attendance as Item 145D	300,000	
M & E Ltd.'s account dated 20/1/2011		12,900,000
Profit pro-data item 145B	5%	645,000
General attendance as item 145C		300,000
Other attendance as item 145		300,000
	13,200,000	14,145,000
Less omissions		13,200,000
Nett addition carried to summary		**945,000**
Lifts installation		
P.C sum as item 1467	20,000,000	

(Continued)

(Continued)

Profit as item 146B 5%	1,000,000	
General attendance as item 146C	250,000	
Other attendance as item 146D	350,000	
Ehi lifts ltd 's account dated 20/3/2013		18,000,000
Profit as item 146B 5%		900,000
General attendance as item 146C		250,000
Other attendance as item 146D		350,000
	21,600,000	19,500,000
Less additions	19,500,000	
Nett omission carried to summary	**₦2,100,000**	

5.7 SUMMARY

This chapter covered measurement of variations and interim valuation, final accounts, daywork and variation account preparation, fluctuation, preparation of prime cost, and provisional sums accounts. The next chapter gives an in-depth description of insurance and bonds as applicable in the construction industry and associated claims.

Exercise

Question 1

Prepare the relevant details of interim valuation No. 1, showing the amount due for payment for a project whose particulars are given below:

Contractor:	A bungalow block of a 4-bedroom flat for Messrs. John, A.Y.
Employer:	Auchi Technical College
Contractor:	HND QS Ltd.
Payment:	Based on the monthly valuation
Contract sum:	₦4,294,983:79K, limit of retention: 5%, completion period: 6 months
Summary of bill of quantities	
A. Preliminaries	192,750:00
B. Measured works	2,797,710:75k
C. PC and provisional sums/provisional items	
i. Electrical installations	320,000:00
ii. Plumbing installations	290,000:00
iii. External works	250,000:00
Contingencies	250,000:00

(Continued)

(Continued)

Value added tax	204,523:04
Contract sum	₦4,294,983:79k

Notes:

1. 25% of the contract sum is paid as a mobilisation fee to be recovered in the first three (3) instalments of payment.
2. The breakdown of the measured work shows that substructure N590,000 is at 100% completion and frames and walls ₦129,500 are at 30% completion, respectively, as at the same time of this interim valuation No. 1.
3. ₦30,000:00 was paid to the utility board as a connection fee.
4. Materials on site are:
   - 35 bags of cement @₦2,300/bag
   - 3 trips of sharp sand @₦9,000/trip
   - 2 trips of granite @₦52,000/trip
   - 1200nos of 150-mm thick hollow blocks @₦120/block

Question 2

Prepare the relevant details of the final account showing the amount due for payment for a project whose particulars are given below:

Contractor: proposed office complex for Dele Friday with an underground car park.

Employer: Isah ventures, payment: Based on monthly valuations, contract sum: ₦105,000,000:00, limit of retention: 5%, completion period: 52 weeks.

Summary of Quantities	
A. Preliminaries	8,100,000:00
B. Measured works	51,426,400:00
C. Nominated sub-contracts	
i. Aluminium component	5,500,000:00
Add for profit 5%	275,000,00:00
Attendance	275,000:00
ii. Plumbing installations	4100,000:00
Add for profit 5%	205,000:00
Attendance	82,000:00
iii. Electrical installations	6100,000:00
Add for profits 5%	305,000:00
Attendance	122,000:00
iv. Lift installations	4500,000:00
Add for profit 5%	2800,000:00
Attendance	225,000:00
v. Suspended ceiling	950,000:00

(*Continued*)

(Continued)

Add for profit 5%	47,000:00
Attendance	23,350:00
vi. Nominated suppliers	
i. Iron mongery	2,800,000:00
Add for profits 5%	140,000:00
ii. Sanitary fittings	1,000,000:00
Add for profits 5%	50,000:00
iii. External works	8,548,750:00
Contingencies	5,000,000:00
Valve added tax	5,000,000:00
Contract sum	N105,000,000:00

Notes:

(1) The contractor could tender for aluminium components and has submitted a quotation of N5,125,000:00, which was the lowest and subsequently accepted.

(2) The following quotations were accepted in respect of items included in the bill of quantities:

Nominated subcontractors:

(a)	Aluminium components	5,125,000:00
(b)	Plumbing installations	4,200,000:00
(c)	Electrical installation	5,700,000:00
(d)	Lift installation	4,300,000:00
(e)	Suspended ceiling	1,010,000:00

Nominated suppliers

(a)	Iron monger	2,650,000:00
(b)	Sanitary fittings	900,000:00

(c) A claim of an extra value of N315,00:00 has been made regarding the increase in the price of iron monger.

(d) Variations and addition works have been valued at N3,515,000:00 as an addition to the contract sum.

(e) The external works have been re-measured and valued N9,125,000:00

(f) Previous payments, including value-added tax, have been made to the value of N90,252,000:00.

Question 3

Based on the extract from the bill of quantities in Appendix A, prepare the valuation accounts for the following architect instructions

Al Nos.	Date issued	Subject
3	10/5/2020	introduce 50-mm-thick precast paving slabs in the soakaway area of the building (area of soakaway 6.0m × 4.5 m wide).
4		Concrete (1:2:4–19 mm Agg) in bed as against concrete (1:3:6–25 mm Agg)

Appendix A

S/No	Items in the Contract Bills	Unit	Qty	Rate N
1.	Excavate topsoil, average 150 mm deep.	m^2		50
2.	Remove topsoil from site.	m^3		450
3.	175 mm bed of broken hardcore to make-up level	m^2		850
4.	50-mm precast concrete paving slab size: 450×450mm laid on hardcore (m/s) and jointed c/m (1:3)	m^2		300
5.	250×100 mm precast concrete coping weathered on top, twice throated, far finished on surfaces, bedded, and joined in cement mortar (1:3)	M	15	750
6.	275×100 mm Ditto	M	12	825
7.	Plain in-situ concrete (1:3:6–25mm Agg) in beds 100-mm thick laid on hardcore	M^3	15	20,000
8.	Reinforced in-situ concrete (1:2:4–25-mm Agg) in suspended roof slabs 125-mm thick	m^3	10	26,000
9.	12-mm-diameter mild steel straight and bent bar reinforcement to BS4449 in suspended slabs	Tonnes	0.75	199,500

Question 4

(i) Define and explain the term "Final Account."
(ii) Itemise and briefly explain the constituents of a standard final account summary.
(iii) State the difference(s) between an interim valuation and a Final Account Summary.

Question 5

(i) Itemise concerning the FMW or JCT condition of the contract, under which circumstances can variation be authorised.
(ii) Define and explain the term "DAYWORK."
(iii) Identify the work activities that are excluded from the fluctuation provisions.

Question 6

With reference to Appendix A, prepare the valuation account for the following architect instructions.

AI Nos.	Date Issued	Subject
8	28/5/2011	275 mm × 125 mm precast concrete coping in Lieu of 250 × 100 mm.

REFERENCES

Ameyaw, E.E., Chan, A.P.C., Owusu-Manu, D.-G. and Coleman, E. 2015. A fuzzy model for evaluating risk impacts on variability between contract sum and final account in government-funded construction projects. *Journal of Facilities Management*, *13*(1), pp. 45–69. https://doi.org/10.1108/JFM-11-2013-0055

Mustaffa, N.E., Nayan, R., Aminuddin, F.A. and Judi, S.S. 2023. Issues encountered in the valuation of interim payment process in construction projects. *Journal of Legal Affairs and Dispute Resolution in Engineering and Construction*, *15*(1), p. 04522047.

Oyegoke, A.S., Powell, R., Ajayi, S., Godawatte, G.A.G.R. and Akenroye, T. 2022. Factors affecting the selection of effective cost control techniques in the UK construction industry. *Journal of Financial Management of Property and Construction*, *27*(2), pp. 141–160.

6 Insurance and Bonds

6.1 INSURANCE

Insurance is developing and spreading in the construction industry because of society's needs and demands. It has reduced the risk for the parties involved in the contract. There is no universal definition of the word "insurance," but it can be described as one party (the insured) accepting to make payments to or for the advantage of the other party (the assured) upon the incidence of certain required actions (Ashworth, 2012). It implies the reasonable financial contribution of many for the advantage of a person who has suffered a loss. Construction activities are not exempt from risk. This is because of the elements involved, such as man and machine. The insurance agreement approach is one practical approach to managing this risk, especially in the construction industry (Odeyinka, 2000). The insurance agreement between these parties is enclosed in an insurance policy document. The consideration is offered by the assured in the form of a premium. Thus, the risk is transferred to insurance companies. Insurance obligations are an important prerequisite in the construction agreement. This is to protect the interests of the parties concerned. Undoubtedly, those advising employers and administering construction agreements have an elementary understanding of the principles of indemnity and insurance. Ramus *et al.* (2011) identified areas in this knowledge that will assist. They are:

(i) Clients are educated about the obligations regarding insurance against risks under a construction contract, hence empowering them to consent to any commitments under the agreement conditions and to include extra spread where necessary.
(ii) The most appropriate insurance clauses (from the different alternative standard clauses) for singular projects are employed, and a satisfactory level of financial protection is stipulated within agreement conditions.
(iii) The insurance strategies for a construction contract comply with the agreed conditions.
(iv) The right methods are utilised in the event of an insurance guarantee related to damage or loss.

However, there are two major types of insurance.

(i) Liability Insurance: This type of insurance offers financial cover for legal liabilities.
(ii) Loss Insurance: This type of insurance covers a loss that falls under the category of insured parties.

DOI: 10.1201/9781032661391-6

All insurance contracts are linked to the common law doctrine of uberrimae fidei. The term "uberrimae fidei" is used to describe an insurance contract "based upon good faith between the parties" (Anifalaja, 2010). Therefore, an agreement of insurance is a legal contract. The assured must ensure full disclosure of every known material fact. A material fact is information that, if revealed, would influence the insurer's decision. In line with the principle of indemnity, insurance should be able to put the insured back in the original position if the risk had not happened. That is, no loss and no gain. The common law doctrine imposes a responsibility on the parties involved to act towards each other with the utmost good faith before and during the contract. This is germane. The construction sector has a weak performance record regarding injury and death, especially among mostly causal staffers during construction activities. This is the worst in many developing countries. Although the overall numbers of deaths continue to decline at a well-supervised construction site with a high level of safety tips daily, accidents and fatalities remain unacceptable, with no evidence of compensation to the people involved. Many factors are attributed to this. In insurance management, the insurer (the insurance company) has a right to represent the insured (the client) to execute the rights and remedies that the insured has against third parties. This is known as "sabrogation." It is a right held by insurance carriers to legally pursue a third party that causes an insurance loss to the insured. This is done to recover the financial claim paid by the individual's insurance firm for losses directly and seek reimbursement from the other party's insurance firm. This insurance mechanism allows the at-fault party's insurer to reimburse the victim's insurance firm.

6.2 INSURANCE POLICIES FOR THE INDUSTRY

The insurance policies for the construction industry offered platforms whereby the contractor could transfer the risk-bearing duty to the insurance firm. These are captured in Clauses 18, 19, 20, 12, and 22. All risks associated with completing the construction activities in compliance with the contract documents are transferred to the insurance firm.

6.2.1 Clause 18—Injury to Persons and Property and Employer's Indemnity

Clause 18.1 states that the contractor indemnifies the employer regarding the personal injury or death of any person for which claims may arise in several ways. The indemnity is limited in two ways. The first is a natural limitation for those persons who could sustain a claim. Thus, in most cases, a trespasser would be unable to claim. Second, in most cases where the employer is not responsible for his act or neglect or by that as any persons for whom he is responsible. This latter term includes persons employed by the employer and would presumably cover the design team. If the contractor wishes to avoid liability under this clause, the onus of proof is effectively upon them. This is because he is the companion at the end of the comments on Clause 18.2. Clause 18.2 demands construction contractors indemnify clients regarding any expenses, risk, loss, claim, or proceedings emerging from damage or harm to property, real or individual.

6.2.2 Clause 19—Insurance Against Injury to Persons and Property

Main contractors are liable to indemnify the client against injury to persons and property. Therefore, the construction contractors are liable for the work. This includes the subcontractors. The contractor should ensure the latter keeps proper insurance to cover the liability. The contractor is only exempt in the case of carelessness caused by the client to persons for whom the client is directly accountable. It is the main contractor's responsibility to indemnify the client against all likely claims. The amount of insurance coverage required is usually inserted in the appendix to the contract conditions (Aniekwu and Okpala, 1988).

6.2.3 Clause 20—Insurance of the Works Against Fire, etc.

The three alternative clauses are provided with emphasis on the third one. This is because the first two are not required and should be deleted. They are as follows:

Clause 20A—Insurance of entirely new work by the construction contractor.
Clause 20B—Insurance of entirely new work by the client.
Clause 20C—Insurance of work of alternation or extension and the existing structures and contents by the client.

Where a construction project consists of two physically distinct portions, the one being new work and the other alternations and extensions, it is quite possible to use the first of the alternatives along with the third, qualifying them as each is applying to their respective part of the work. The relevant clause for an agreement should be inserted in the agreement particulars. Clauses 20A and 20B are alike, except that in Clause 20A, the contractor takes out the policy for all-risks insurance.

6.2.4 Insurance Option A—Clause 20A

The main contractor should remove and keep a joint name policy for all risk insurances in this option. This is for full reinstatement value. Also included is the percentage stated in the appendix as a contract. Where the main contractor neglects to give satisfactory insurance cover regarding the agreement, the client may guarantee against the dangers portrayed. The client cannot stretch the policy to cover other risks yet can subtract the premiums from the sum owed to the construction contractor under interim payments. If loss or damage occurs on an insured risk, the following steps should be taken:

(i) The contractor should send a written notice to the Architect and the client. The notice should capture the details of the extent of damage.
(ii) No record should be taken of the damage or loss when computing installation payments due to the construction contractor.
(iii) The main contractor must authorise the insurers to pay amounts due under the insurance claim to the client.

(iv) The main contractor is paid for the re-work by the client in instalments and is certified by the Architect. These certificates should be released at the normal time of interim certification.

6.2.5 Insurance Option B—Clause 20B

In Clause 20B, the client must provide insurance for the work. The procedure is like Clause 20A, except regarding payment to the contractor. If the client fails to maintain an insurer, the construction contractor can take out joint names cover regarding the default, and the cost adds up to the contract sum.

6.2.6 Insurance Option C—Clause 20C

This alternative stands with Clause 20B but against Clause 20A by placing the same contractual risk and its insurance with the employer. This clause is utilised when construction includes work on an existing structure, for example, refurbishment, fixing, fitting-out, and extensions. There are two separate issues to manage here:

- Insurance of the existing structure and its subject matter
- Insurance of the fresh/latest works

The client must assemble a joint name policy regarding the existing building and its subject matter. The conditions for insurance of the works and the processes and provisions concerning the occasion of loss or damage are like those under Insurance Option B.

6.3 PROFESSIONAL INDEMNITY INSURANCE

Professional indemnity insurance is not a new concept, but the concept is new in many developing countries. As the name suggests, this is intended to cover those who provide expert service. An example of such insurance is regularly connected with construction professionals, such as Architects, Quantity Surveyors, and Engineers. For instance, some professional institutions, such as the Nigerian Institute of Quantity Surveyors, demand that their members convey professional indemnity insurance if they offer an expert service under their registered status. Current popular procurement systems have shown that construction is fully involved in the design and financial process. In some cases, the client does not even engage the design team. It pursues that where a contracting organisation executes design work or offers expert guidance as a normal part of their work, they must have professional indemnity insurance. The client or his guides should examine this fact. The reason for a professional indemnity policy request cannot be overemphasised, for example, the default in exercising sensible ability in designing the work or selecting materials and components. It is expected that if a contractor sub-let part of the work to a sub-contractor, the main contractor is expected to keep professional indemnity insurance as legally dependable to the client for the competence of that design and other advice. This insurance policy is every year capable and will have a financial limit forced upon the level of indemnity it gives. In practice, although not always applicable, professionals must take professional indemnity insurance for 12 or 15 years after completing the construction project. It is hard for the employer to implement this condition, although it can be a very expensive mistake for any expert to overlook it.

6.4 INHERENT DEFECTS INSURANCE

This is a new concept of insurance, though not popular in African countries. An inherence defect is a defect that exists. Still, it remains unobserved before the date of practical completion but later reveals itself through actual physical damage, which may have to be fairly discovered earlier. A claim in this type of policy is difficult, and often resorts to the courts to recover damages. The origin of the defect could be shoddy supervision, structural failure, poor design, poor workmanship, etc. The difficulties facing insurers in Nigeria and preventing the orderly use of inherent defect insurance are as follows:

(i) Inadequate statistics
(ii) Inadequately experienced underwriters
(iii) The potential cost to the family to underwrite properly
(iv) Weak demand
(v) The poor image of the Nigerian construction sector with lax production and skill levels. This industry ought to be the second-highest employer of labour after the agricultural sector.

The crusaders of inherent defect insurance argue that latent defect insurance assists in promoting buildability and predictability. They also believe that construction projects with inherent defects that insurance covers will benefit the following:

(i) Construction team's professional indemnity insurance may not be necessary.
(ii) The possible conflict is mitigated.
(iii) There is more satisfaction for the project team.
(iv) Less time and money are spent arguing about contract conditions and warranties.
(v) Innovation is reinvigorated.
(vi) Everyone can focus on getting the real design and construction correctly.
(vii) It can be a foremost benefit when negotiating a sale or letting.

6.5 BOND

A bond may be understood as a guarantee of performance. A construction bond is a form of guarantee, not insurance (Taylor, 2009). It is a written agreement in which the surety guarantees that the principal would fulfill its obligations to the oblige (Boswall, 2010). It is a three-party agreement between a surety, principal, and oblige. This implies that the surety is committing its financial resources to guarantee a construction contractor's performance. This is a cost to the construction project and, by extension, to the client. The guarantor (an insurance firm) is a surety for the construction contractor at a cost in case of default. This becomes a cost to a client following the failure of a contract during a construction project. Also, the disappointment of a subcontractor can mean a conceivably benefit making contract bringing in a loss for a contractor. Bonds aim to assure performance and give a monetary reward in case of disappointment. The bonds utilised on the construction projects are:

(i) **Advanced payment bond**
If the employer agrees to give an advance payment (mobilisation fee) to the construction contractor, the employer requests this type of bond. The contract condition would influence this decision. It is known as a labour and material bond. The bond offers security for the principal's suppliers and subcontractors if the contractor fails to pay them (Boswall, 2010). The advance payment bond is equal to the sum of the mobilisation fee and is for 60 days, or as stipulated in the agreement.

(ii) **Bond for payment of off-site material**
There might be events to buy materials and products ahead of time and store them off-site. The merchandise and materials being referred to may be exorbitant, delicate, or both, and, in this manner, it is improper to store them on site. The construction contractor applies for a payment instalment and a bond to take care of the expense of the materials or merchandise should there be harm or loss. The activity of this bond resembles the advance payment bond. The surety's liability is to reward the client for the amount he may have paid to the contractor for the listed materials off-site. The client must state the maximum value the surety might be capable of. The surety's liability finishes on the date all the listed items have been delivered to the site or by a specified longstop date, whichever happens first.

(iii) **Retention bond**
The retention bond favours the client because they (the owners) often take over projects before making completion payments to the contractors. It is a form of cash retention used to protect the employer/client from a contractor's insolvency and ensure the finished work is done properly. A retention bond is normally carried out previously or on the date of ownership and is given to the client. Retention, customarily at 3% to 5%, is subtracted from every interim payment made to the contractor and held by the client until practical completion. This might allow unfaithful clients to evade the responsibility of a complete payment. The bond assures reimbursement to the beneficiary in the event of the contractor's non-performance after the completion of the contract. Other common types of bonds utilised in the construction industry are listed below.

(iv) **Performance bond**
A performance bond is pertinent among construction bonds. It is required to ensure the performance of the contract signed during the work and secure that projects are completed to the quality, cost, and time stipulated in the agreement. It is usually 10% of the value of the contract sum, as stipulated by the Nigerian Public Procurement Act 2007. In Australia, only performance bonds are used for construction projects (Australian Constructors Association, 2009). As the name suggests, a performance bond is planned to

guarantee the acceptable performance of a party to an agreement. This type of bond is mostly used in the construction industry. The client does request a performance bond from the contractor because it acts as a guarantee of the performance of a contract. Also, the contractor does the same as the sub-contractor for the same reason.

(v) **Tender bond**

An employer might require a bond to guarantee contractors who express enthusiasm in submitting a bid for a construction project are bona fide. This should be distinct from tender deposits (provided by a party to the bidding process as against a third party). Both approaches encourage construction contractors to honour their bids by entering the tendered contract. The bond fund might be utilised if a contractor fails to submit a bid or enter into an agreement after being chosen. The range is between 1% and 5% of the tender sum. If the contractor defaults to honour a winning bid, the surety (insurance firm, in this instance, a third party) will pay the owner and seek recovery from the defaulting contractor.

(vi) **Payment bond**

This type of bond guarantees the responsibility to make payments. Contractors might demand them to ensure a client's responsibility to make payments or by sub-contractors to guarantee payment from a contractor. Some finance institutions request this document to finance construction projects or release loans to contractors seeking construction project financing.

6.6 THE USE OF BONDS

There is no defined list of circumstances where a bond should be utilised in the construction industry. Therefore, a choice needs to be reached on each construction project whether to use or not to use a bond. The following are likely circumstances where a bond may be suitable:

(i) To protect the interests of a "one-off" developer.
(ii) In instances that involved a new or untested contractor in a project.
(iii) Tender bond is needed in nominated sub-contracts.
(iv) To ensure that the risks inherent in a construction project are covered.
(v) To ensure a bidder does not withdraw a tender (tender bond).
(vi) To ensure the construction contractor complies with the contract agreement (performance bond).
(vii) To ensure the construction contractor pays back the mobilisation fee (advance payment bond).
(viii) To ensure payment of retention fees after the defect liability period (retention bond).

The argument for or against the use of bonds depends on the perspective of the individual and the situation being addressed. It is expected that construction practitioners would have observed the proper selection and appointment of construction projects before being awarded to contractors and sub-contractors. In this instance, acquiring bonds is unnecessary. For these practitioners, the requirement to provide protection in the form of bonds will add to the cost of a construction project. Therefore, a systematic approach is demanded in considering the demand for bonds.

6.7 CLAIMS

There are three main contractor claims. They are as follows:

(i) **Common law claims**
This kind of claim arises from causes that are not within the agreement's express terms.

(ii) **Ex-gratia claims**
Ex-gratia claims have no legitimate premise but are claims the contractor believes the client has an ethical responsibility to fulfil. For instance, if the contractor has truly undervalued an item(s) whose quality has increased significantly. If not considered, any variation causes him a significant loss. The client is not liable to meet such "hardship claims" yet might be set up. This is on the grounds of regular justice or to assist the contractor where something else is needed to avoid the contractor being forced into liquidation. If this last option occurs, this may not be the best option for the project.

(iii) **Contractual claims**
This type of claim is frequent in construction projects. The following may lead to contractual claims:
- Fluctuations
- Variations
- Extension of time
- Loss/or expense due to matters affecting the regular progress of the work

6.7.1 Preparation and Submission of Claims

The operation of the claim procedure involves the following processes:

(i) **Identification of claim**
The first stage of the preparation of the claim is the identification of the claim. The contractor should identify the contract conditions and incidents that caused the claim. The impact of the incidents on the construction project and contractor should be well documented.

(ii) **Notification of claim**
Under the terms of the agreement, the contractor is obliged to inform the Architect in writing of any event that may result in a claim. This is necessary once it becomes apparent that such an occurrence will cause delay or loss to the contractor. The contractor may lose his right to a claim should he fail to notify the Architect of his intention to claim within a reasonable time.

(iii) **Evaluation of claim**
It is the Architect's responsibility to award the contractor an extension of time or payment for losses and expenses of such proportion as the Architect

considers honest and realistic. In practice, the contractor normally prepares an evaluation that he feels will satisfy his loss for the Architect to consider, upon which the Architect can base his award.

(iv) Submission of claim

Once the contractor has evaluated the claim, he will make a formal submission to the Architect for consideration. The claim submission is a detailed document prepared by the contractor that argues his case and shows details of the evaluation. The submission document will normally include the following:

- Contract particulars are background to the case.
- Claim particulars identify the cause(s) of the claim and an argument for the case.
- The evaluation of the claim represents the contractor's detailed calculations of delay and loss suffered.
- Appendices to supporting documents.

(v) Negotiation and settlement of claim

Once the Architect has considered the contractor's claim submission, the Architect may give the contractor time to argue the case through a meeting before finally awarding the claim.

(vi) The evaluation and submission of claims

The evaluation and submission of claims may be a long and detailed process. This may involve a contribution from design team members, including the Quantity Surveyor. The evaluation and submission comprise all functions involved in assessing the effect of claim occurrences and assessing the loss suffered. The process will involve detailed consideration of any or all of the following information.

- Correspondence
- Site meeting minutes
- Architect's instruction(s)
- Labour allocation sheets
- Site diary
- Daily weather reports
- Receipt of drawing schedule
- Tender build-up
- Plant records
- Day work record
- Programme
- Cost information
- Cost reports

The following are types of construction claims that the contractor may submit:

(a) Delay claims

Delay claims are claims where the contractor attempts to establish that factors outside his control caused the root cause of an interruption in the completion of the work. The contractor is eligible for an extension of time under the contract. Claims of this nature involve the development of an argument relating to the cause of the delay and the contractor's assessment of the duration of the delay caused by such factors. The delay extension will be assessed by a logical consideration of the factors relating to the case by the construction consultant.

(b) Claims for loss and expense

Claims for loss and expense are of three types:

(i) Disruption claims: This is where the contractor wishes to claim the financial effect of inefficiency introduced into his administration of the project due to its occurrence. This entitles him to claim. Such claims may relate to such aspects as:

- Access/method: Due to claim occurrence, the contractor's work method becomes less efficient, more expensive, or even both.
- Acceleration: Where the contractor is forced to speed the progress of the works, this may accommodate a claim manifestation, thus creating extra expenditure on resources.
- Sequencing: Where a claim manifestation affects the contractor's competence due to significant sequencing hurdles.

(ii) Prolongation claims

The contract duration is extended because of claim occurrences and causing additional contractor cost. Items of additional cost that will be incurred because of a prolongation claim include:

- Preliminaries: Since the cost of many preliminary items is directly time-related, for example, staff salaries, extended rental costs on accommodation, scaffolding, equipment, and facilities.
- Climatic costs: Costs incurred because of work executed at a less favourable time of the year.
- Extended attendance: Additional costs of attending to sub-contractors and suppliers.
- Overheads: The contractor may argue that he has been involved in additional overhead costs because of the continuation of the contract.

(iii) Direct loss

The contractor may be able to identify direct losses because of the claim occurrence. The contractor will be entitled to recover this claim if they can justify the loss directly related to the claim. Suppose the contractor attempts to profit from claims or recover losses not caused by claim occurrences. In that case, the contractor will likely receive little sympathy from the Architect and the Project Quantity Surveyor. The contractor should honestly justify his claim before he submits it. The evaluation of loss and expense claims can be made on various losses. The contractor's ultimate objective to recover should adopt a logical evaluation that can be understood by those who are to appraise the claim; if not, it may end up as an exercise in futility.

6.8 SUMMARY

This chapter covered an in-depth description of insurance and bonds applicable to the construction industry and associated claims. Also, the chapter identifies Clauses

18, 19, and 20 as applying to the industry and their implications. The next chapter gives an in-depth description of the financial statement and progress regarding construction projects. This includes liquidated and ascertained damages.

REFERENCES

Aniekwu, A.N. and Okpala, D.C. 1988. Contractual arrangements and the performance of the Nigerian construction industry (the structural component). *Construction Management and Economics*, *6*(1), pp. 3–11.

Anifalaje, K. 2019. Statutory reform of the Doctrine of Uberrimae Fidei in insurance law: A comparative review. *Journal of African Law*, *63*(2), pp. 251–279.

Ashworth, A. 2012. *Contractual Procedures in the Construction Industry*, 6th ed. Harlow: Pearson.

Australian Constructors Association. 2009. Bonding issues faced by construction companies in Australia, a report by KPMG Corporate Finance (Aust) Pty Ltd.

Boswall, R.G. 2010. Construction bonds guide. *Clark Wilson LLP*. Available at: www.cwilson.com (accessed 23 June 2011).

Odeyinka, H.A. 2000. An evaluation of the use of insurance in managing construction risks. *Construction Management and Economics*, *18*(5), pp. 519–524.

Ramus, J., Birchall, S. and Griffiths, P. 2011. *Contract Practice for Surveyors*, 4th ed. New York: Taylor and Francis Groups.

Taylor, D.K. 2009. Construction payment and performance bonds myths and reality. *Journal of Property Management*, pp. 18–19.

7 Financial Management

7.1 FINANCIAL STATEMENT AND PROGRESS REPORT

Good construction management practice requires regular financial reporting to monitor a construction project. The report should be produced on a monthly or quarterly basis. A financial report for a construction contract should contain part or all of the following:

i. Initial tender figure and anticipated profit
ii. Projected figures at completion for value and profit
iii. Current payment application by the construction contractor
iv. Current certified value
v. Amendments to the certified valuation
vi. Costs to date and the accounting period in question
vii. Cash received to date, retention deducted, and certified sums unpaid

Normally, after the contract has been set, the Architect should assemble a gathering of all concerned. This can occur in the Architect's office or that of the contractor. This is to enhance facilitation and make it known to one another. Also, it will allow for the opportunity to decide the process to be employed at subsequent gatherings. The Architect should usually attend this meeting as the chairman and other construction consultants, for example, Quantity Surveyor, Engineers, agents of the contractor, and possibly nominated sub-contractors and suppliers if awarded already. The Architect must organise regular site meetings. The intricacy of the project will impact the rate of these meetings. It is expected that the contractor should submit a progress report to the consultant and client. As part of their responsibilities, the Architect should forward the progress report to the client (employer).

The Architect gets weekly or monthly reports from the clerk of works covering such issues as identified by Seeley (1997):

i. Men are used in the various work sections.
ii. Weather conditions and information regarding the time lost.
iii. Principal deliveries of materials and lacks.
iv. Records of the plant on site.
v. Details of contract drawings and other documents are needed.
vi. General progress compared with the programme.
vii. Any other issues affecting the activity of the contract.
viii. Reference numbers of Architect's instructions received.
ix. True reflection of work done to date.

The progress report is a treasured method for educating the Architect about progress on the site. This will provide a helpful wellspring of reference should disputes emerge along this line. A dispute is inevitable in contract administration. The progress report

DOI: 10.1201/9781032661391-7

is regularly arranged from the diary kept by the clerk of works. Therefore, a financial statement is a summary valuation of the total cost of all work during preparation (Ashworth *et al.*, 2013). This reflects the financial implications of all instructions and changes determinable on a project up to the time of the statement. It is desirable to keep an accurate account of the contract price. It notices ordered and proposed instructions from the initial cost to the final price. It is pertinent to keep records of factors affecting the project's contract as quickly as possible, even before the instruction. Also, it helps carry the project owner through the unfolding financial commitment up to the end of the project. This allows an interval in the likely completion cost of the project. The report should be produced, ideally every month. The content of a financial statement depends on who is preparing it. The one highlighted earlier (i–vii) is an example of the one prepared by the contractor's QS for his boss/employer. An example below is a typical financial statement from a Consultant Quantity Surveyor to the project owner.

i. Initial contract sum
ii. Client-revised requirement
iii. Fluctuation
iv. Variation (other revised requirements and adjustments)
v. Estimated (Anticipated) final cost
vi. Net: Cost over-run/Anticipated savings

Example of Financial Statement

Pace Consultant
Chartered Quantity Surveyors and Project Managers

15, Igbe Road, Benin City — **Ref: 20/08/2022Ref: 20/08/2022**

Contract: Block of classroom	Financial Statement: Nr. 6
Contract sum	12,500,000
Less: Contingencies	250,000
	12,250,000
Add: Variations (schedule attached)	250,000
Add: Claims (schedule attached)	75,000
	12,575,000
Add: Allowance for contingencies	25,000
Estimated final cost (fluctuations excluded)	12,600,000
Add: Allowance for fluctuations	230,000
Estimated total final cost	**12,830,000**

Date: August 20, 2022 — Sign..................................

7.2 LIQUIDATED AND ASCERTAINED DAMAGES

These are damages paid by the contractors for non-completion of work on time, in line with Clause 23.1. Clause 24 allows the claiming of liquidated and ascertained damages from the contractor if he neglects to complete the construction works as planned for insufficient reasons. The sum inserted in the appendix will ordinarily be computed flatly every week or at intervals. Note that specific consideration ought to be taken to guarantee the sensibility of the sum inserted so that the courts do not construe the amount as a penalty and, in this way, unenforceable. This should be checked completely at the early stage before the endorsement of the contract by the parties involved.

7.2.1 Certificate of Architect

The starting point is for the Architect to issue a certificate stating that the completion date has been overrun. No date for issuing this is given, but it is reasonable that it should follow strictly the completion date to avoid uncertainty. Either party may, of course, dispute this certificate. Two essential elements are encompassed within the expression "the completion date." They are Clause 23.1 (an initially defined completion date) and Clause 25 or Clause 33 (provision for the extension of time). Some of the causes allowed in Clause 25 are matters that would otherwise enable the construction contractor to substitute any realistic finish date for the fixed date of the appendix. These are breaches of contract and within the power of the client or his Architect.

7.2.1.1 Payment or Allowance of Liquidated Damages

The construction contractor is not responsible for liquidated damages pending the matter of the certificate under the proceeding clause. Damages include any duration between the finish date recognised and the date for practical completion, which may or may not have passed at the date of the certificate. The client has a responsibility to set regarding the claims for damages, but there is no obligation on the contractor's part to pay them automatically. The design team members, including the Architect and Quantity Surveyor, need the authority to adjust interim or final certificates to account for them. Such an adjustment is not mentioned in any part of Clause 30.

However, it is prudent for the Architect to advise the employer that an adjustment has not been made when issuing a certificate. The client's written notification to the construction contractor should be given on the day of the issue of the final certificate. However, this may not pay sums increasingly or, more likely, "allow" them as deductions from payments made by the client. The client may deduct the losses from the balance in the final certificate if necessary. If this balance is not enough or in the contractor's favour, the employer may take the usual steps to recover the money as a debt.

The completion date will not be finally fixed under Clause 25.3.3. The Architect's certificate must be treated first under this Clause if the implied sequence of the contract events is followed. But if the completion date is finally fixed later, any excess damages are to be returned to the contractor. A provision for liquidated damages defines the contractor's risk and saves the employer the trouble and expense of proving their actual loss in litigation. It is not an issue, even if the loss is difficult to assess. For example, Lord Halsbury LC remarked in the Clydebank Engineering Co. Ltd. case that "the very reason the parties agreed to such a stipulation was that, sometimes, the nature of the damage was such that proof would be extremely difficult, complex, and expensive." Therefore, an agreed damages clause of this type is valid and implementable. The sum is a sincere pre-valuation of the loss that is probable to be caused to the employer by late completion, or a lower sum. Case ref: Clydebank Engineering and Shipbuilding Co. Ltd. vs. Don Jose Ramos Izquiredo Y Castaneda (1905). In the aforementioned case, shipbuilders agreed to construct four torpedo-boat destroyers for the Spanish government at an agreed price. Also, it was agreed that the delivery would be within the specified period. The contract provided that liquidated damages for delayed delivery "should be 500 pounds a week for each vessel not delivered by the contractors in the agreed time." The vessels were delivered late. The Spanish government paid the full price but claimed 500 pounds each week for late delivery from the shipbuilders. The House of Lords held that the claim was enforceable and summarily dismissed the shipbuilder's argument "that there could be no measure of damages in the case of a warship which had no commercial value at all."

7.2.2 Lord Halsbury LC Said

It is a strange and somewhat strong affirmation to state that the damages could easily be ascertained regarding the commercial ship. However, the same principle could not be applied to the warship, as it earned nothing. The duplication of a country's warships might mean intense harm, although it is probably not easy to ascertain the sum. It seems hopeless to argue in such a debate. It is necessary to express the statement to show how preposterous it is. These principles apply with equal force. In the construction industry, in some cases, such as those of educational buildings, ecclesiastical buildings, and housing association projects, contractors sometimes argue that the amount fixed for liquidated damages is such as to render it a penalty. This is either because it is said—there is no loss—or because the likely loss is impossible to estimate. The converse is the case. In such circumstances, it is customary and sensible to use a simple formula approximating the detailed analysis of all individual costs. This formula is as follows:

i. Assume that at the anticipated completion date, 80% of the total capital cost of the scheme (including fees) will have been assuming an interest rate of 12%.
ii. Administrative costs are assessed as 2.75% of the contract sum per year.

The net result of (i) and (ii) of the directly above formula reduces to 0.237% of the contract sum weekly. Such a figure may underrepresent the losses, but such a calculation is a fair and reasonable approach to use. There is another calculation recommended. Under this, it is suggested that 30% of the total capital cost of the scheme is multiplied by the current interest rate and then divided by 52 to give a weekly figure. Such formulae are useful where the actual loss is difficult to assess. But in the commercial contract, the client and his advisers should estimate the possible loss when the agreement is made. If the figure is a sincere pre-valuation of the loss as reasonably envisaged, then the court will not interfere with it. This becomes the figure recoverable, whether the genuine loss is greater or less.

Liquidated damages should be distinguished from penalties, but the distinction can be challenging to appreciate. Lord Cranworth LC acknowledges this in the case of Ranger vs. Great Western Railway Co. (1854). The difference between a penalty and a fixed sum is that the standard amount of damages is too well proven to be called into question. A penalty clause is unenforceable. A clause will be a penalty and not a sincere pre-valuation of loss (1) if the amount stipulated is exaggerated concerning the highest possible loss that can be experienced, or (2) as a general rule if a single amount is payable concerning various breaches with widely differing consequences. This second principle is inapplicable to standard form contracts in the construction industry because the standard liquidated damages clauses are limited to the breach of late completion. The principles the courts apply to determine whether a clause is a provision for liquidated damages or a penalty were summarised by Lord Dunedin in Dunlop Ltd vs. New Garage Co. Ltd (1915) in a speech regarded as a classic exposition of the position. The case law supports the following propositions:

(i) Though the parties to an agreement who utilise the words penalty or liquidated damages may prima facie be intended to mean what they say, the statement utilised is not decisive. The responsibility is on the court to see if the payment inserted is, in honesty, a penalty or liquidated damages.
(ii) The quintessence of a penalty is a payment of a sum inserted as a deterrent to the culpable party. The quintessence of liquidated damage is a sincere. covenanted pre-valuation of damage.
(iii) Whether an amount is inserted as a penalty or liquidated damages is an issue to be decided based on the terms and circumstances of each contract. This is decided during the period of making the contracts, but not at the hour of default.
(iv) To help this construction project, several tests have been recommended, which, if appropriate to the subject matter, may demonstrate accommodating or convincing behaviour.

Examples:

(a) It will be considered unacceptable in sum, in contrast with the greatest loss, which could be proved to have followed from the default breach.
(b) It will be held to be a penalty if the default comprises only not paying a certain amount and the sum inserted is a sum greater than the sum that should have been paid.
(c) When a single lump sum is inserted by way of compensation, the assumption is that it is a penalty. The significance of the damages will not be counted this time and may occasion serious or insignificant damages.
(d) It is no obstruction to the amount inserted being a sincere pre-appraisal of damage that the consequence of the breach is such as to make precise pre-appraisal almost impracticable. This is only the circumstance when it is most likely that the pre-assessed damage was a genuine deal between the parties.

It is a defence to a claim for liquidated damage to establish that the stipulated amount is a penalty. Furst *et al.* (2012) point out, he may either rely on his claim for the penalty, in which case he cannot recover more than the actual loss which he proves up to the amount of the penalty, or he can ignore the penalty and sue for unliquidated damages: Watts, Watts & Co. Ltd. Vs. Mitsui & Co. Ltd.

If the sum is found to be a penalty, this is not a complete bar to the employer's claim. They are merely left to prove their expenses and losses, which they recover through unliquidated damages. Liquidated damages clauses are caught by the provisions of the misnamed Unfair Contract Terms Acts 1997, which were designed to limit the use of exclusion or exemption clauses.

7.2.2.1 Liquidated Damages Position Under JCT 1980 Forms

Clause 24.1 makes it a conditional guide to the client's right to liquidated damages. This enables the Architect to issue a certificate that the construction contractor has failed to complete the work by completion. The certificate is a simple statement of fact and is merely confirmatory. It is issued after the Architect has considered all claims for extension of time under Clause 25, which have been presented by the contractor when the current completion date is reached. The Architect should issue a fresh certificate if he alters the completion date in his review under Clause 25.3.3. Clause 24.2.1 introduces more condition patterns and clarifies that the deduction of liquidated damages is a discretionary right of the client. Liquidated damages are to be paid by the contractor if the employer gives written notice to the construction contractor of his intent to claim or subtract liquidated damages. The Architect should do this before the issue of the final certificate. The wording clarifies that the client may claim the total or part of the liquidated damages. The contractor is only bound to pay or allow "to the client, the whole or part of such as may be specified in writing by the client." If the client fails to give notice before the final certificate is issued, he loses his right to liquidated damages and is left to claim at common law.

If the Architect reviews the completion date under Clause 25.3.3 after revising contract development within 12 weeks of the date of practical completion and fixes a later completion date, Clause 25.2.2 gives the contractor the right to be reimbursed for any liquidated damages that the client deducted for the period up to the later completion date. Two final points should be noted. This is to clear up common misconceptions. First, the JCT 1980 liquidated damages are concerned with liquidated damages for the interruption in completion. The client's eligibility to claim unliquidated damages for other breaches of agreement by way of an action at common law remains intact. Second, no liquidated damages are payable if the figure is not inserted for liquidated damages in the appendix, whether through oversight or otherwise. In such a case, the client's remedy for the delay in completion would be a court decision for unliquidated damages at the law court, with the necessity of providing a loss. Equally, should no completion date be stated in the appendix, no liquidated damage will be paid (Kemp vs. Rose (1858)). This is because the date for when liquidated damages for the project may begin to run is missing.

7.3 EXTENSION OF TIME

This means an extended period is granted to the contractor to complete work because of a delay caused by the client. Extension of time provisions and liquidated damages clauses are firmly connected. And disappointment by the Architect's right to exercise power to extend time, for which the delay is caused by the employer (or for which the employer is liable for land), relieves the contractor from his obligation to pay liquidated damages (Dodd vs. Churton, 1897). The time for completion may become large. The influence of the operation of this Clause is to lessen or remove the force of Clause 24. This benefits the client and the contractor: first, the client by keeping the right to liquidated damages alive; second, the contractor, by relieving him of their payment instantly. The Clause is also drafted to avoid some of the pitfalls of granting an extension that have arisen in the past.

In several places, the Architect should fix an amended completion date, usually in writing and sometimes by notifying the contractor rather than issuing a certificate. The effect of the extension is to modify one of the terms of the contract, and both parties should be notified since some of the causes of the extension are matters of the Architect's default. He is here acting as both judge and defendant.

Three definitions are given for Clause 25. "Completion Date" includes any amended date arising from an extension, while "delay," notice, or extension includes any further such happenings. "Relevant Event" is the most significant in that only events delineated in Clause 25.4 can lead to an extension of time.

7.3.1 Notice by Contractor of Delay to Progress

The intention for obtaining an extension of time beyond the original or amended date for completion is for the contractor to give notice "forthwith." Such action and its promptness may permit alleviating action to be taken under Clause 25.3.4. In his initial notice, the contractor should set out "the material circumstances," which include the cause or causes of delay and, by implication, should cover any uncertainties and interacting matters. This should be followed by the claim for an extension, as indicated under Clause 25.1. The contractor is to "estimate" the "expected" delay, if possible, in his notice, but otherwise as soon as possible, and he is to do this separately for every relevant event. This is important for segregating current events. The present delay may also overlap with some earlier delay leading to an extension or even other circumstances arising after the notice. This information will help the Architect check the contractor's delay estimate and look for means of reducing the delay under Clause 25.3.4.

Lastly, the contractor should give further notice to the Architect covering changes in the estimate or delay and the associated effects. These notices may be "reasonably necessary" at the contractor's option or "as the Architect may reasonably demand" if the information is inadequate for him to keep the situation under review. However, at these three stages, the contractor must send copies of the documents to any nominated sub-contractor referred to in the original notice. He should include any other firm which becomes involved as the situation develops. This serves to bring Clause 35.14 regarding extensions for nominated firms, and it is important that all affected nominated sub-contractors are mentioned and their representations are brought forward. It may be a process or a sub-contractor that has caused the delay under the relevant event of Clause 25.4.7. While a nominated supplier may cause a delay under the same provision or be affected by any other delay here, he is not entitled to receive any copy documents and need not be mentioned in them. This reflects that the nominated supplier is not involved in the site activity.

7.3.2 Fixing Completion Date

Experience shows that the Architect may have difficulty making a firm early estimate of the delay. The Architect is under a general duty to give an extension as soon as possible so that the contractor is confident about the measures to take to meet the completion date. If the Architect fails, the Architect may lose the power to grant an extension and thus lose the client's right to liquidated damages. The Architect, in some cases, may not be able to give a fair extension until after the previously fixed date. This may be close to completion, and the Clause recognises this in its staged procedures. The Architect must exercise "his opinion" on what is fair and reasonable here and later in adjusting this estimated extension. While he must be fair, it may be noted that, regarding Clause 25.3.3, he cannot finally revise his opinion towards a shorter extension except based on further suitable variations. Therefore, caution may be advisable at this stage. The Architect extension is to be in writing and state the relevant events and variations based on the decision. This will allow the contractor to object if anything relevant is excluded from the approval. When such an extension has been granted, the contractor should amend his master programme under the optional Clause 5.3.1.2 if it applies.

Once the Architect has acted under this clause during the contract period but not before, he may fix an earlier completion date than the modified date. This is because the Architect has, in the interim, instructed omission variations that are quite independent of any relevant events. The final fixing of the completion date should be inserted within 12 weeks of practical completion. This constitutes a review of the previous date, whether original or interim, and allows for a later, earlier, or unchanged final date to be fixed. The strict wording of the clause makes the three options mutually exclusive and appears to allow the Architect to act only once after practical completion. There may well be causes tending to both retard and advance the date. This is only reasonable for the Architect to take both into account when finally fixing the completion date. There are cases where the Architect takes the initiative to make a second attempt at fixing the completion date. It would appear open to the party potentially disadvantaged by such action to object successfully. There would be a strong reason for one of the parties to concede a second revised date or face arbitration or other action because the Architect's decision was unreasonable.

Four provisions are given here, of which the first two may be taken together in Clause 25.3.4. In part, these two underline the contractor's duties under Clause 23.1 "regularly and diligently to proceed." In the case of delays, he is to use his best endeavours and take avoiding action on his initiative. The measures include progressing material orders, sub-contractor's performance, and that of the Architect. Even if a delay occurs, he must try to prevent the delay from affecting the completion date. The contractor needs to cooperate with the Architect to reduce the effects of delays. It would be reasonable to limit the extension granted by allowing regard to the effect of the contractor's ameliorating action. The third provision (Clause 25.3.4.3 or 25.3.5) is that the Architect must notify "every nominated sub-contractor of every completion date he fixes." But he does not have to give reasons, as in the case of the contractor. Finally, regarding Clauses 25.3.4.4 or 25.3.6, Architect decision is not allowed to modify the completion date earlier than the date in the contract appendix. The extensive omission variations may even be without regard to the question of acceleration.

7.4 RELEVANT EVENTS

The causes of delay here are called "Relevant Events," which might lead to an extension of time and lead to claims under Clause 26.1. This is for additional payment or determination by the contractor under Clause 28.1.3. They are discussed in full below, but for convenience of comparison, the causes given in the respective clauses are tabulated in Table 7.1. Also, Clause 33.1.3 governs the position where war damage occurs. There are some reservations over Clauses 25.4.3, 25.4.7, and 25.4.10. They all arise either by the action or inaction of the client or the Architect or through circumstances outside the power of either party to the construction agreement.

TABLE 7.1
Causes of redress under Clause 25, with similar causes under Clauses 26 and 28 of JCT

	Relief or Redress Available to the Contractor		
Cause of Delay	**Extension Under Clause 25**	**Determination Under Clause 28**	**Payment Under Clause 26**
Exceptional weather	Yes	No	No
Strikes, lockouts	Yes	No	No
Delay by nominated firms	Yes	No	No
Delay by statutory bodies	Yes	No	No
Inability to obtain labour, goods, or material	Yes	No	No
Exercise of statutory control	Yes	No	No
Action on the discovery of antiquities	Yes	No	No
Force majeure	Yes	Yes	No
Fire, flood, storm, etc.	Yes	Yes	No
Civil commotion	Yes	Yes	No
Inadequate ingress/egress	Yes	No	Yes
Architect's variation orders	Yes	Yes	Yes
Discrepancies leading to Architect's instructions	Yes	Yes	Yes
Postponement on Architect's instructions	Yes	Yes	Yes
Delay in obtaining drawings, instruction	Yes	Yes	Yes
Delay by others engaged by the employer	Yes	Yes	Yes
Opening-up and testing	Yes	Yes	Yes

- The determination following the fire and flood is partially covered by Clause 22C.
- Clause 28 contains other grounds for determination besides extensions as an alternative remedy. Others appear in Clauses 32 (hostilities) and 33 (war damage).
- Clause 34 (Antiquities) also provides the contractor with the right to recover losses not otherwise covered by the contract.

Force majeure is not a phrase native to English law, and its meaning is imprecise but wider than the act of God (an interesting theological point). Force majeure refers to exceptional issues outside the power of either party (refer to Clause 25.4.1). Exceptionally advise weather conditions. The first word is important; weather may be extraordinary in its degree, its timing, or its perseverance. Local meteorological records should be consulted to establish a norm (refer to Clause 25.4.2). Loss or damage caused by Clause 22 perils. This is in respect of the world itself. The contingencies may result from the contractor's default, as the discussion is beyond the scope of this topic. However, where Clause 22A applies, Clause 20.2 may carry the day against the present Clause (refer to Clause 25.4.3).

Civil commotion (and the like). The provision extends to all the processes preceding the site operations, at least for the goods or materials directly needed for the work. Civil commotion so far as it causes damages to the project is already covered in Clause 25.4.3 (refer to Clause 25.4.4). Architect Instructions (in certain cases). This covers all instructions over the following matters (refer to Clause 25.4.5):

a. Resolution of discrepancies or divergences, perhaps with additional work or with a resultant waiting time.
b. Variation and the expenditure of provisional sums are routine physical changes and developments.
c. Postponement, not involving physical change
d. Action (and inaction) on the discovery of antiquities.
e. Nominated sub-contractors and suppliers, particularly but not only over nomination, determination, and re-nomination.

However, in cases (a)–(c), as in Clause 25.4.3, there is no exclusion of the contractor's negligence. Clause 28.1.2.4 excludes this, and is the default. Also covered by this clause are instructions for opening it up for inspection or testing. But the liability for the resultant delay is linked to the liability for the cost of making goods under Clause 8.3. This includes not having received important directives, drawings, and the like. In these cases, "due time" will be necessary, for instance, to obtain materials or goods through the normal channels and also sufficient margin to assimilate all the information required for proper production planning, often a prolonged matter for today's complex buildings. The wording "neither unreasonably distant from nor unreasonably close to" is intended to guard against situations in which the contractor may time his requests for information so that he may embarrass the Architect and himself have a lever for a claim. The Architect may be delayed in issuing information because he has yet to receive design information, which is the responsibility of a nominated firm. While the contractor will receive an extension here, the employer may seek compensation from the nominated firm under Agreement NSc/2, Agreement NSc/2a, or the Warranty Agreement Schedule 3 of the Nominated Supplier form of tender, as appropriate (refer to Clause 25.4.6).

This is a delay on the part of nominated sub-contractors or nominated suppliers. This clause, in isolation, relieves both the contractor and the nominated firms of responsibility quite unreasonably, considering that they have a direct contractual connection. Still, the client has such a relationship with the contractor only. The clause has no bearing on domestic sub-contractors or other suppliers. If the contractor has found a way to prevent or lessen delay, he is granted an extension, preventing damages. Furthermore, the defaulting nominated organisation does not need to meet the contractor's claim to be reimbursed for such damages. He remains responsible for other delays of disturbance costs, and the client is left with no redress. The defence for the nominated organization is to build up its default, and the contractor attempted his best to surmount or prevent it (refer to Clause 25.4.7).

Clause 29 deals with the overall question of work on-site not included in the contract and distinguishes two categories. Clause 29.1 covers work about which the contract bills give enough information for the contractor to assess its effect on his overall contract obligations. While Clause 29.2 covers the alternative category of work about which information is inadequate or non-existent. From experience, it is advisable to state information in the contract bills to satisfy Clause 29.1. To state that the contractor is to allow for any effects on his programme so that the question of an extension does not come into the debate (Refer to Clause 25.4.8).

Statutory powers directly affect the availability of labour, materials, or energy. The reference to any statutory power widely causes this but is also closely restricted so that not every action by or on behalf of the government qualifies. First, it is restricted to an "exercise" after that date of tender, and so the contractor should take account of any earlier exercise with delayed effect on the works. Second, it is restricted to an exercise by the UK government. It does not embrace, say, a national corporation's marketing actions unless they are constrained by suitably direct government intervention. It must directly impede him by restricting, preventing, or delaying the contractor. In the case of the supplier occurrence, the contractor must turn to the next clause to support his case (refer to Clause 25.4.9). This clause has the protection that the contractor should "reasonably have foreseen and provided for all likely problems of this kind via local enquiry and the like." The wording should indicate how far the contractor should go before asking for an extension or what criteria the Architect should apply before granting one (refer to Clause 25.4.10).

7.5 CLAUSE 25.4.11

A statutory undertaker is executing or defaulting to carry out work. This parallels Clause 25.4.7 and is necessary because Clause 6.3 removes the bodies concerned from the scope of the nomination system where work is solely in pursuance of (their) statutory duties. Delay will be classified under Clause 25.4.7, where a nominated body executes work as part of the building services. A construction contractor cannot seek an extension (refer to Clause 25.4.11). This may lead to the client's collapse into offering entrance to or way out of the project site. This is distinguished from failure to give ownership of the site under clause 23.1, which is a fundamental breach. The present clause relates to ownership and control of the client and not part of the site, which does pass into the contractor's possession. This is at some "due time" and may occur only briefly. The clause provides that it may arise out of some provision in the contract bills or in the contract drawings. It may also be contingent upon the contractor giving notice to the Architect, and failure to do so may defeat a claim of extension (refer to Clause 25.4.12).

7.6 SUMMARY

This chapter covered financial statements and progress reports regarding construction projects. Also, the chapter presented liquidated and ascertained damages and how payments are prepared for management with work examples. The next chapter gives an in-depth description of conflicts in construction contracts and their root causes.

REFERENCES

Ashworth, A., Hogg, K. and Higgs, C. 2013. *Willis's Practice and Procedure for the Quantity Surveyor*. West Sussex: John Wiley & Sons.

Furst, S., Ramsey, V., Hannaford, S., Williamson, A., Keating, D. and Uff, J. 2012. *Keating on Construction Contracts*. London: Sweet & Maxwell.

Seeley, I.H. 1997. *Quantity Surveying Practice*. London: Macmillan.

8 Conflicts in Construction Contracts

8.1 CONFLICTS IN CONSTRUCTION: BACKGROUND

The construction industry is significant in all nations since it adds to the quality of life. Construction activities involve a variety of events, depending on the size and type of construction projects, professionals, and trade skills required. Construction contracts can fluctuate from work with a few thousand naira to major schemes costing several million naira and some over a billion naira. Construction contracts in the past consisted of a page or two-page document. However, because of the development and transformation taking place in the construction industry, a complex and onerous set of conditions try to cover every possibility. An attempt to achieve this has created loopholes that the legal profession can feast upon. On a different platform, participants in the construction industry concur that construction agreements need to become less confrontational and emphasise the optimistic demands of the project. Team working is one of the possible ways to achieve this goal rather than fault-finding individual backs all the time. The Nigerian construction industry offers customers great flexibility. This is jealous of many parts of Africa and developing countries in other parts of the world. The construction industry is also known as the built environment. This is complex by the presence of stakeholders with an interest in the completed project. The stakeholders include the client, also known as an employer (sometimes the end-users), the funding parties, the developers, the planning authority, the policy regulators, and the public at large, among others. These stakeholders hold the built environment in high esteem, especially regarding society's economic, environmental, and aesthetic roles.

A school of thought declares that the high number of interested parties within the built environment process enhances the catalyst for conflict in the industry. Although this is arguable, facts over the years show that disputes in construction are common. Therefore, many see conflict in the construction sector as a normal occurrence. There are various forms of construction conflicts. This ranges from the two-party differences over the interpretation of a construction agreement clause to the public disagreement conflicts common occurrences within the construction industry. Conflicts are harmful both to the industry and to employers. The costs of solving conflicts are often high, and the period can sometimes be long. The influence of conflicts on construction projects can jeopardise the mission of the stakeholders concerned. It can sometimes be this disastrous. Since the construction industry is risky, conflicts are likely to arise even under the best circumstances. Even in the presence of elimination of possible potentials, it will still arise. Ashworth *et al.* (2013) identified some of the main areas where conflicts might occur:

DOI: 10.1201/9781032661391-8

General

1. The confrontational nature of building/construction contracts
2. Weak communication between the stakeholders involved
3. The proliferation of contract and warranty types
4. Fragmentation in the construction industry
5. Tendering procedures and processes

Clients

1. Weak instructions
2. Changes and variation conditions
3. Changes to the standard conditions of the agreement
4. Interference in the contractual responsibilities of the contract administrator
5. Delay in payments

Consultants

1. Design shortfalls
2. Inadequate skill and experience
3. Delay and deficient information
4. Inadequate coordination
5. Undecided delegation of obligations

Contractors

1. Poor site management
2. Inadequate planning and programming
3. Poor workmanship
4. Disputes with sub-contractors
5. Delay in payments to sub-contractors
6. Weak coordination among sub-contractors

Sub-contractors

1. Incompatibility of subcontract conditions with the main agreement
2. Default in following and adopting agreed procedures
3. Bad workmanship

Manufacturers and Suppliers

1. Failed to define performance and objectives
2. Failed to perform

The construction industry embraces many crafts and professions. The industry is, therefore, prone to conflict. Many parties are involved, each with its own values, benefits, interests, education, and needs. What, then, is conflict? The chapter classified conflict into four parts:

a) Military war: War between opponents, particularly a drawn-out, severe, sporadic battle.
b) Difference: A divergence or conflict between thoughts, standards, or individuals.
c) Psychology Mental Struggle: A mental state coming about because of frequent insensible opponents between concurrent yet incongruent wants, needs, drives, or motivations.
d) Literature, not tension: Disagreement between or among characters or powers in an artistic work that shapes or propels the activity of the plot.

"Conflict" and "dispute" are two separate terms. Conflict exists wherever there is a mismatch of intrigue and pandemic in this manner. Conflict can be overseen to the point of thwarting a dispute resulting from the conflict. A dispute is related to distinct justiciable matters. Disputes demand solutions. This implies that they can be controlled. The process of dispute resolution lends itself to third-party intercession. The source of dispute is when the needs of one or two parties still need to be met. The solution can resolve the specific dispute at hand. Conflicts are often centred on more problematic issues. A conflict often hinges on high-stakes moral values that are difficult to negotiate amicably. New procedures, for example, risk management and partnering, are being developed to tackle the causes of conflict. Still, Nigeria's built industry is viewed by many as claims-oriented and a platform for conflict and dispute. Henceforth, there is a need to oversee conflict to circumvent the following:

1. To avoid the increase in crises
2. To prevent the worsening of relationships between the parties
3. To circumvent undesirable influences on the timing and quality of the product because of the conflict
4. To mitigate the costs of conflict resolution

8.2 CAUSES OF CONSTRUCTION CONFLICT

The major causes of construction conflict are wide-ranging and varied. The following are the major causes of conflict, as highlighted:

i. **Misunderstandings**
Misunderstandings generally happen due to a lack of communication. There is no exchange but speech; both people talk, but no one hears or listens. When a manager sends their subordinates off to execute a task without clarifying what they (managers) anticipate achieving, either that or the subordinates set off without behaving and listening mindfully to their instructions.
ii. **Sensitivity**
There are days when a person's mood will rapidly change, which may scare other partners, friends, and co-workers. This mood may swing and influence

the person's manner, patience, and tone. This could affect their judgement when they should consider the needs of others.

iii. **Values**
Values vary between individuals, experts, and skills. Professional experts in the construction industry have codes of practice that denote minimum standards of conduct. Satisfying these guidelines may result in discipline or expulsion from their professional institute. When there is a crisis, these values can cause an inward clash.

iv. **Interests**
Some individuals have impracticable hopes. The employer wants speed and, at the same time, a quality building at the lowest cost. The contractor may require additional time, realistic quality, and maximum cost to make an optimal profit. In some extreme cases, their interests differ, but they need not act as such.

v. **Personalities**
Feelings play a major role in the conflict. The inability to handle stress may cause conflict. A person's self-esteem may also lead to conflict. However, those engaged in conflict often adopt a hard-combative position. They do not want people to evaluate them as weak for adopting an appealing method.

vi. **Environment**
Studies have shown that our environment may influence who we become in society. Persons in different parts of the world become habituated to their climate and may find other climatic environments repressive. Therefore, our environments affect our speaking, dynamics, geography, infant experiences, childhood, and religion. It has been argued that our ideas, beliefs, and interests are formed to a large extent by our environment, our childhood, and our education.

vii. **Education**
Education levels, both formal and informal, can impact construction conflicts. Education can be described as the method that moves us from an innate response to inspired or inventive behaviour. Our capability to reason and generate our own ideas would be limited if not for our acquired education. Also, education makes the communication of ideas to others very easy. The level of education or awareness of the situation may determine the degree of conflict.

viii. **Experience**
In some cases, a contractor knows that a solution is not feasible, but this knowledge is not known to the employer. While a contractor believes a solution will work, the employer rebuffs the idea. Our experience can either be limiting or liberating. It largely depends on our outlook. Psychologists tell us that every experience, good and bad, is deposited in our memory and will be retrieved subconsciously when we want to make decisions. Our past experiences affect the number of ideas we have.

ix. **Ideas**
If anything is unavoidable in the construction contract, it is ideas. The construction industry is one industry where ideas are germane. This can create a problem if not well utilised. Individualistic ideas of approaching a problem

can cause conflict if one side's belief or experience is different from the other side's belief or experience. Conflicts of ideas often arise and are connected to conflicting beliefs.

x. **Change**
Change in a construction contract is inevitable. Changes to design plans, deadlines, payment dates, etc. can cause conflict if poorly managed. This may lead to doubt about what the finished construction project will include and misperceptions of what the different contractors were expected to incorporate in their total sums.

xi. **Delay**
Some clauses inserted in the construction contract show high anticipation of failure. They are assumed to have been written in anticipation of one party's failure to comply with the agreement signed. But the intention is to protect the other party should there be a delay.

xii. **Quality**
Some parties often need to define quality consistently. This has created many issues in the construction industry. Except quality is satisfactorily defined, it stays as a subjective challenge. Therefore, the definitions of quality should be accurate and proper to the price we are ready to offer.

xiii. **Money**
Money has a relationship with quality. A subcontractor may misinterpret your requests and quote a lower price than other sub-contractors. Conflict is likely to occur when the mistake is revealed to him.

xiv. **Time**
They say time is money in the world circle, including in the construction industry. Time is significant in projects because the completion dates of the milestones are already prepared before the commencement of the project. There is a clause to remedy liquidated damages if there is a default. So, the construction process is time-controlled from inception to possession.

8.3 CONFLICT REDUCTION IN CONSTRUCTION PROJECTS

The construction industry could become a platform for conflict if the risk of an event deteriorating into conflict is not mitigated. There are 12 constructive steps that can be taken when attempting to manage conflicts. They are as follows:

8.3.1 Step 1: Communicate with Precision

Misrepresentation or misunderstanding because of inadequate structured communication. This can be a factor in causing conflict. This is because many works take place within groups or organisations. However, we communicate on three levels. These three levels are significant in well-structured communication.

- Content (the spoken words)
- Tone (the way we say the words)
- Visual (the body language)

For communication to be precise and pretty good, we should avoid ambiguity and speak plainly. We should always state our views clearly and be ready to speak honestly and fairly, even if it feels uncomfortable.

8.3.2 Step 2: Listen and Consider Attentively

Communication is bidirectional. So, one should be willing to hear what the other person says. It is necessary to allow the other side to hear their views just as you do. Also, you should read thoughtfully, look prudently at what is written, and identify the document's subject matter and tone.

8.3.3 Step 3: Think Before Speaking

Experience has shown that many react quickly to what we see or hear without a second thought about the message our response sends to others. One possible way to mitigate conflicts in the built environment is to process what we intend to say before saying it.

8.3.4 Step 4: Take Time to Build Relationships

Building a relationship within a team to achieve the goal cannot be overemphasised. This will help reduce the likelihood of sensitive conflicts. People choose not to oppose those in whom they have invested time and emotion.

8.3.5 Step 5: Be Honest in Your Dealings with Others

Distrust has been identified as one of the major causes of conflicts. There is a need to separate honesty from being blunt. The truth can be presented diplomatically without harming other parties involved. It is easier if there is a friendly connection with the other party.

8.3.6 Step 6: Do Not Dispute Trivial Matters

Ignoring trivial matters during the discussion will help to reduce conflict. I had an experience during one of my site meetings in 2003 when one of the consultants said, "you are a common HND holder." I ignored the statement and made my presentation with supporting evidence. At the end of that meeting, the client sacked the Architect because his arguments were based on demeaning his opponent and were baseless. Therefore, avoiding unwarranted arguments will sustain goodwill and lessen the likelihood of additional conflict.

8.3.7 Step 7: Look for Common Ground

Parties in conflict must search for common ground to find answers to issues rather than debating against each other's suggestions. Finding a common platform will inspire a feeling of teamwork.

8.3.8 Step 8: Recognise and Avoid Prejudice

Differences in the construction industry cannot be avoided because the parties involved are more than one. In the construction field, we will meet different people. The approach and manner we attend to them say a lot about us. We all deserve to be treated equally regardless of our educational background, tribe, sex, religion, colour, etc. Prejudicing others on the grounds of any of these will enhance bad feelings.

8.3.9 Step 9: Express Your Understanding

There is a need for parties involved in a contract to understand what their challenger is saying in any conversation. After carefully listening to what your challenger has said, you may recite back to him the facts as you have grasped them. This approach would save time and mitigate the possibility of misunderstanding.

8.3.10 Step 10: Control Your Emotions

Emotions and personal feelings should be controlled in dealing with construction matters. If this is handled with wisdom, it may avoid an issue. A good example is "you HND holder," in a site meeting. I was able to control my emotions, discard his comment, and make my presentation with a systematic approach. At the end of that meeting, we got the victory. If I had reacted, the meeting that day would have been distorted.

8.3.11 Step 11: Apologise Gracefully if You Are Wrong

Construction practitioners should have "I am sorry" in their dictionary. Many of us find it difficult to say sorry. We are human beings and can make errors, even in judgement, and misread others. Some practitioners believe that they are always accurate, but no one is. An apology for a wrong done is likely to be restitution for the error or wrong.

8.3.12 Step 12: Accept Apologies Gracefully If Others Are in the Wrong

For someone to acknowledge that he has erred, admitting and saying sorry takes courage. It is necessary to state here that it takes great courage to apologise. Therefore, it is expected that we should respect this courage by gracefully acknowledging sincere regrets to promote good behaviour in the future.

8.4 DETERMINATION OF CONTRACT

In common contract forms, some provisions allow either party to terminate the contract. The decision to terminate a contract should be justified with evidence that a breach of contract has happened. The word "determination" is a term that has been misused within and outside the industry. It is a term that explains deciding whether to terminate an agreement or not between a contractor and employer (client). The action contemplated by Clause 25 FMW, under which the employer can determine, and by Clause 26, where the contractor determines, is referred to several times as determining the contractor's employment under this agreement. This should be distinguished from determining the agreement. The two clauses on determination in this present condition set out the procedure to be followed in the determination of employment and the constitutional rights and responsibilities of the two parties until the final settlement. These two clauses determine only the employment of the contractor. Except in Clause 25.2, there is no mention of reversing the procedure and reinstating the contractor's employment. In the event of reconciliation, the parties must make arrangements and agree on terms.

8.4.1 Determination by Employer

8.4.1.1 Clause 25.1 Determination by Employer—Clause 25

Four ways of defaulting are given, any of which would be a breach of contract. Therefore, the contractual grounds for determination by the employer are the following:

25.1.1 "Wholly suspends" represents a complete withdrawal of manpower from the site; the present equipment alone will not be enough defence since nothing is being carried out.

25.1.2 "Fails to proceed regularly and delight" represents a fitful approach, and falling behind the programme without cause will put the contractor in breach of Clause 21.1.

25.1.3 Refusal or wilful neglect in removing defective work or the like must be such as to have a serious effect. A less drastic remedy here is to employ other persons to carry out the removal in question. Where it is impracticable for such people to carry out the work while in possession, owing to its extent or serious nature, the determination alternative is available here.

24.1.4 Assignment or subletting in themselves would not be causes for determining at common law. They are under Clause 17 and can therefore be brought within the scope of this present clause.

8.4.1.2 Clause 25.2 Contractor Becomes Bankrupt

The provisions of bankruptcy and company law are beyond the scope of this book. However, there is some doubt as to whether the present clause would be valid if tested in court. The clause will likely be upheld where there is a receiver or manager on behalf of debenture holders. Where there is actual insolvency, it would be seen that the basic principle of the law would prevail. This implies that the trustee could choose whether to reject the agreement in the interests of the creditors. The weakness of the clause, in principle, is that it endeavours to put one creditor, the employer, in an advantageous position compared with the rest. The clause differs from 27.1 by making determination automatic when bankruptcy or the like occurs. The clause also mentions the possibility of an agreed-upon reinstatement of the contractor's employment. This would be possible in any case, and the clause gives no terms or procedures. Whether it is wise to reinstate the contractor will depend on the stage of the work, among other things. If the contractor's employment is reinstated, there is always the risk to the employer that the trustee or the like may exercise his continuing right to disclaim later if the financial state is more advantageous to the creditors. The clause has its snags but will often be accepted as the basis for settlement.

8.4.1.3 Clause 25.3 Not Used

This clause appears only in the local authorities' contract edition, heading "corruption." It is beyond the scope of this book.

25.1.4 The procedure for determining this in each case given requires only two registered or recorded letters. The basic timetable is as follows: (i) The Architect writes specifying the default; (ii) a waiting period of 14 days is allowed for the contractor to cease his default; and (iii) if he does not, then the employer may determine within a further ten days. The employer may determine within the next ten days even if the contractor ceases his default immediately after the 14-day waiting period. If the employer does not act within his period, he will have lost his right to determine if the contractor has resumed work. If the contractor continues to default beyond the 14 days, it will then seem reasonable for 10 days not to run until the contractor stops defaulting. This will allow the client to determine at any time during the extended default period or for 10 days after. However, it would be possible for the contractor, if negligent or unscrupulous, to cease default within 14 days and then resume later. If the Architect must serve a fresh notice on the contractor and initiate again, it would be possible for the contractor to be continually going into default and coming out again, yet for the employer to be unable to determine. It is therefore provided that if the contractor repeats a default for which he has already received an Architect's notice, the employer may determine within 10 days and without any waiting period whether the contractor will continue in default. For this to take effect, the default must repeat the original default to fall within the same four categories. A default in another of the four categories would require a fresh procedure.

Finally, it is possible to see that both parties could abuse this clause by playing close to its letter, and therefore, two safeguards are given. For the employer's benefits, a determination is devoid of discrimination against any other rights and remedies, a phase commented on under Clause 24.6. The other is for the contractor's protection. The determination shall not be brought about unreasonably or veraciously, that is, on the grounds of a quite minor but strict default or perhaps by a too harsh operation in a case.

8.4.1.4 Clause 25.4 Determination of Employment of Contractor—Rights and Duties of Client and Contractor

The scene procedures are given to finish the work and come to a settlement, no matter the determination cause. Their effectiveness will be affected by the considerations mentioned under Clause 25.2 where this is relevant. They do not apply if the contractor's employment is reinstated.

25.4.1 Employment of others. This is the clear sequel to the preceding steps. No restriction is placed on the employer regarding the type of contract or contract he may enter upon severe completion. Much will depend on how piecemeal the balance of the work is. The employer does not need the approval of the contractor or any trustee for their commercial dealings. Still, he will be expected to use an economy consistent with progress. Some of these will have become the employer's property in the case of materials and goods. Whatever is used will come into the final reckoning under Clause 25.4.4. However, in the case of plants and the like, the employer has no right of ownership under these conditions but only a right to free use. He will have to account for these items under Clause 25.4.3. He cannot retain hired items under Clause 24.4.3 if the hirers agree.

25.4.2 Suppliers and sub-contractors. The scope is not restricted to nominated firms only but is perfectly general. The first main provision under Clause 25.4.2.1 is that if the contractor remains inactive, the employer has an option within 14 days of determination to ask for an assignment of agreements. This is of value where continuity is required. The second main provision under Clause 25.4.2.2 is that the employer may pay such firms any money not yet paid by the contractor, except where there has been insolvency. This is general regarding the timing of work concerning the determination. This covers the practical situation where firms work past determination due to uncertainty about what has happened to the contractor.

25.4.3 Removal of plants and the like. The employer will be liable for any loss or injury outside of fair wear and tear if the contractor removes items when required. If the contractor defaults on removal, the clause relieves the client of this responsibility. It does put the client responsible for passing the net sales results over to the contractor, in other words, to the contractor's credit. This is in keeping the ownership of the plant and the like at no time passing to the employer, and its value cannot be set directly against any contractor debt to the employer under Clause 25.4.4.

25.4.4 Final settlement. It will take over the account of the items on the client's side:

a. The employer will report loss directly and/or damage due to the determination. A most likely head claim here would be delayed in completing the work. A basis like that of liquidated damages is likely suitable. Although Clause 22 as such would be suspended by the determination since Clause 22 rests on Clause 23.
b. Expenses incurred by the employer. These would include retention monies held on nominated sub-contractors before determination, temporary measures following determination, and the amount paid for work and materials to complete the construction project. These later would, in turn, include additional professional fees and insurance. These expenses could only be questioned if it could be shown that the employer had proceeded with a gross lack of commercial reasonableness.
c. Sum received by the contractor before the date of determination. This is the sum of all interim certificates met by the employer. It is not required that the exact value of work before determination shall be calculated.

The "monies" paid may include payments concerning nominated subcontractors that the contractor has a default to pass on, and the client may or may not have paid directly.

These three items are to be set against "the total sum which would have been payable or due completion" against a hypothetical final account prepared since the contractor's prices or other terms throughout. The resultant difference is a debt payable by one party to the other as appropriate after the issue of a certificate.

8.4.2 Determination by Contractor

26.1 This clause presents all the causes for determination by the contractor. Insolvency, in this case, does not warrant separate procedures and is included here. But several causes are also included in all the cases in which the contractor may "forth worth" determine his employment. In Clauses 26.1.1 and 26.1.3, "forth worth" operates at the end of a waiting period, but only under the former does the contractor need to indicate that he intends to determine. All of which is quite harsh.

26.1.1 The employer does not pay for any certificates. This clause operates after 14 days or more of default, followed by a seven-day warning period. The reference is "the amount properly due to the contractor on any certificate."

26.1.2 The client obstructs the release of the certificate. The employer and the Architect will prefer to be in direct association throughout the progress of the work. At no point must this operate to influence the impartiality of the Architect, and this may be a fine point of distinction. It is only the employer's action that is envisaged and must be demonstrated, not the inefficiency or obstructiveness of one of his professional advisers.

26.1.3 The carrying out of the uncompleted work is suspended. This may come into play for several causes, which are named in Clause 23 as 'Relevant Events' and under which they are discussed individually. Some of them also occur as "matters" in Clause 24. Determination is the drastic remedy for them and, unlike the other remedies, can only be implemented after a "continuous period" of delay. The delay duration is to be inserted in the appendix, but in default of any insertion, it will be one month by default. Only four out of the seven causes of suspension call for comment here. One is Clause 26.1.c.ii with its reference to Clause 20 perils, which are events requiring a delay of three months according to the appendix unless some other period is stated. The reference to Clause 20 perils introduces whichever of the variants of Clause 20 applies. If the contractor determines when Clause 20 applies, the insurance money will pass to the employer through the joint coverage of the policy. Clause 20B needs no commitment. Clause 20C also contains its provisions for determination to meet its circumstances of frustration and scale of work rather than delay, which is in view here.

The second that calls for comment is in Clause 26.1.c.ii, where "civil commotion" is mentioned alone, and its direct effect is not qualified as in Clause 23.4.4. It is already covered by Clause 26.1.c.ii, regarding the physical damage to the works. The third such cause is in Clause 26.1.c.iv, where negligence and default of the contractor may lead to action by the Architect, and here both are excluded. Clauses 26.1.c.iv to 26.1.c.vii have a clear effect on deterring the employer of the Architect from dilatoriness or delay. The Architect should act promptly even where the contractor is at fault under Clause 26.1.c. iv. The fourth cause to mention is a delay in the carrying out of work, separate from the agreement under Clause 26.1.c.vi. This is different from simply performing work, as in Clause 23.4.8.1, where the problem is mentioned of the distinction between work under Clause 27.1 given in the agreement bills and that, under Clause 27.2, not so given. This ceases to be a problem here because delay is defined. There may be delays due to unexpected work, but the contractor should secure an agreement about this before consenting to it happening. If this is likely to lead to a prolonged suspension, he might be able to withhold his consent.

8.4.2.1 Clause 26.2 Determination of Employment by Contractor

This clause governs several particulars but states that what is provided is without prejudice to two overriding considerations. The first is "the accrued constitutional rights or remedies if any party," which might exist as contra-claims to modify the amounts otherwise payable under Clause 26.2B. This may be due to statutory fees, liquidated damages, insurance premiums, claims settlement, direct payments to nominated sub-contractors, or other causes provided throughout the conditions. The second consideration covers matters of injury and indemnity arising under Clause 18. It is provided that the influence of the clause is extended to cover the contractor's activities on-site, as he is still required to carry them out after determination. In this case, the term "without prejudice" is uncertain.

However, at determination, the contractor stops carrying out the work as Clause 18 describes it. So, his indemnity stops; nothing is left to which the present clause may stand "without prejudice." Clause 26.2.9 treats Clause 18 as a thing of the past and gives its definition of liability, which it seems must stand, given the inapplicability of "without prejudice."

26.2.a The contractor's right to remove all is also made an obligation so that the employer may reassure possession of the site free of all temporary items and materials mentioned. Many of these might be of great use to the employer if he is going to complete the work, but unless he makes an attractive enough offer to the contractor outside the contract, they will go. The contractor is to remove all his materials and goods. Regarding the word, "his" must reasonably be interpreted as applying to items the contractor owns, or at least to which his title is no worse than the employer. If the client has paid the contractor, the employer will prima facie have a better title but may need an adequate one, as mentioned under Clause 14. Removal is made subject to Clause 26.2.b.iv, under which the employer purchases items that are "his" rather than the contractor. Once the employer has paid the contractor, he cannot remove it. They are no longer his items. But until the employer has paid, the contractor may remove what is "his," even if the employer is bound to pay for it in due course, punitive though having to pay for such action might be.

26.2.b The contractor is paid for work carried out under Clauses 26.2.b.i and 26.2.b.ii as in the interim certificate. This will depend on the measured rate from the contract bills and the rates derived. Clause 26.2.b.iii refers to loss and expense other than that due to the determination itself and therefore caused earlier, as in Clause 24. Furthermore, the cost of materials or goods covered under Clause 26.2.b.iii includes "those for which the contractor lawfully bound to make payment." This will often include items not delivered by the determination date. Clause 26.2.b.vi covers the loss and expense due to determination and, curiously enough, specifically mentions losses caused to nominated sub-contractors.

8.5 SUMMARY

This chapter covered conflicts in construction contracts and their causes. Also, the chapter presents how conflict can be reduced and the procedure for determining the contract and employer.

REFERENCE

Ashworth, A., Hogg, K. and Higgs, C. 2013. *Willis's Practice and Procedure for the Quantity Surveyor*. West Sussex: John Wiley & Sons.

Index

Note: Page numbers in *italics* indicate a figure, and page numbers in **bold** indicate a table on the corresponding page.

Printed in the United States
by Baker & Taylor Publisher Services